28/23 sci £-

CW00521990

fo

by David Houghton

Totally Chlorine Free   Sustainable Forests

Telephone 0845 345 0400 for a free copy of our Publications Catalogue.

Cover Design: Petra Clifford
Typeset: Creativebyte
Proofreading and indexing: Alan Thatcher
Printed in China through: World Print
Acknowledgements: Met Office, Frank Singleton, MCA, UKHO, MarineCall, Wetter Zentrale, MDL Marinas.

David Houghton is a world renowned Meteorologist, an expert on the subject, having written Weather Forecasts first in 1968 and updated annually, as well as countless other books, journals and articles on the subject.

David was Weather Adviser for over 30 years for a number of different sailing campaigns including the British Olympic Team, Admiral's Cup, Whitbread teams, the British, New Zealand and French America's Cup campaigns.

He has also taken part, as Meteorologist, in 9 British Olympic campaigns and, as both Meteorologist and sailor, in the 1981 Fastnet race.

David Houghton's knowledge of weather forecasting, world weather patterns and climatology is vast to say the least.

Foreword

We Brits love to talk about the weather, and if you're a boater... well that just makes things even worse. Whatever kind of boating you are into, the weather really can make or mar your trip. In the most extreme case it can be the difference between having fun and being in danger. Despite all of this, many of us still struggle to grasp the fundamentals of weather forecasting, and this is where this book really comes into its own.

David Houghton has put together a nice simple guide which is easy to understand and clearly laid out. It enables the reader to gain a far better grasp of what the weather is going to do.

None of us can control the weather, but if you have a firm grasp of what is going to happen, then you have a far better chance of picking the right day to get out on the water. To my mind, it is also extremely rewarding to understand things like *why* the wind shifts at a certain time of the day and what causes a sea breeze and when it kicks in. This kind of knowledge taps into that elusive quality, seamanship, all boaters constantly strive for.

RYA Weather Forecasts will provide an invaluable companion to any boater, providing you with the nous to get the most out of your boating, whatever the weather.

Rod Carr, OBE

Contents

Your Weather Forecast

It is just as important to follow the weather as it is to know the tides. Using both wind and tide to advantage by intelligent attention to the weather and weather forecast can increase enormously one's sense of achievement. And no longer is it necessary to scratch around for every scrap of information gleaned from broadcasts of uncertain origin; the network of weather observations from accredited stations in northwest Europe is better than ever, and accessible in real time.

From the BBC's Shipping Forecast to the on-board automatic reception provided by the Navtex kit the sailors' needs for weather information are being met. Pride of place must go to the BBC Shipping Forecast as the world's longest running weather forecast service, still largely in its original and well proven format. Even the names given to the sea areas around the UK have achieved legendary status, and in addition to the thousands who listen because their livelihood depends on the information given, there is a growing core of devotees for whom listening to the Shipping Forecast is an integral part of their way of life.

This book:
- identifies the most useful and reliable sources of weather information and forecasts available to sailors in European waters
- provides basic guidance on the effects of coasts and islands on the wind and on the development of sea and land breezes
- unwraps the language of weather forecasts including words used to describe wind, visibility, barometric trends etc.

Abbreviations

GMDSS   Global Maritime Distress and Safety System

SOLAS   Safety of Life at Sea

MRCC   Maritime Rescue Co-ordination Centre

MRSC   Maritime Rescue Co-ordination Sub-Centre

AWS   Automatic Weather Station

MSI   Maritime Safety Information

GW   Gale Warning

SWW   Strong Wind Warning

MCA   Maritime & Coastguard Agency

MSI Broadcasts

Broadcast alerts for all MSI broadcasts are transmitted on VHF Channel 16 and 2182 kHz. VHF Channels used are 10, 23, 73, 84 and 86.

The first wave of transmissions commences at 0710 and 1910, the second wave at 0730 and 1930, the third wave at 0750 and 1950, then finally MRCC Clyde commencing at 0810 and 2210.

GMDSS - INMARSAT-C

The Global Maritime Distress and Safety System (GMDSS), of which NAVTEX is an integral part, provides for the worldwide automatic receipt on board of navigational and weather information in English transmitted by satellite (Inmarsat) via Standard-C. GMDSS also encompasses radio, telex and telephony broadcasts for coastal waters. Warnings and forecasts for METAREA I, the northeast Atlantic north of 48° 27'N, provided by the Met Office are broadcast from the BT Inmarsat station at Goonhilly. METAREA II covering the northeast Atlantic south of 48° 27'N, and METAREA III, the Mediterranean, are the responsibility of the French and Spanish Met Services respectively. Sea areas used in the broadcast from Corsen for METAREA II are shown on page 60. The times of the main weather bulletins, all of them in English, for the areas are:-

METAREA I   – 0930 and 2030 GMT

METAREA II   – 0900 and 2100 GMT

METAREA III   – 1000 and 2200 GMT

For more detailed information see pages 41-42.

NAVTEX

Navtex is an international navigational telex service broadcasting safety related messages in English on 518kHz. Dedicated receiving equipment is readily available and not expensive. The simplest version comprises a receiver tuned to the frequency, a small printer and a chip which controls what is received and printed. All messages are prefixed by a 4-character 'word'. The first character identifies the transmitting station, the second the category of message and the remainder are two serial numbers. Details will be found on pages 41-42. This facility to receive gale warnings and marine weather forecasts automatically is particularly valuable for anyone cruising offshore. Twice daily bulletins cover all sea areas around UK and Western Europe.

A national Navtex service on 490kHz has been developed in co-operation with France and provides more detailed Maritime Safety Information (MSI) covering all coastal waters of North West Europe, along with a three to ten day outlook (see page 42).

MF

MSI is broadcast by the following MRCCs via MF on the designated frequencies shown below, after announcement on 2182 kHz.

		Routine A Twice a day as shown:	Routine D Twice a day as shown[3]:
Clyde	1883 kHz	0810, 2010	0210, 1410
Shetland	1770 kHz	0710, 1910	0110, 1310
Stornoway	1743 kHz	0710, 1910	0110, 1310
Falmouth	2226 kHz	0710, 1910	0110, 1310
Aberdeen	2226 kHz	0730, 1930	0130, 1330
Humber	2226 kHz	0750, 1950	0150, 1350

Three day forecasts, primarily for the benefit of fishing fleets in the North Sea, Northern and Western Isles and the South Western Approaches, are broadcast from 1st October until 31st March. The three day forecast is part of the standard MSI schedule of broadcasts made twice daily on MF.

VHF

MSI is broadcast by MRCCs on either VHF Channels 23, 84 or 86, and exceptionally Channel 10, following an initial brief announcement on Channel 16. The initial announcement indicates the working channel on which the broadcast is made. Routines are broadcast at three-hourly intervals by the MRCCs, starting at the times shown in the following table.

Routine A

Full MSI broadcast, including new Inshore Waters Forecast and Outlook, Gale Warnings[1], Shipping Forecast, and the 3 Day Fisherman's Forecast when and where appropriate.

Routine B

New Inshore Waters Forecast plus Gale Warnings.

Routine C

Repetition of Inshore Waters Forecast and Gale Warnings as per previous Schedule A or B Broadcast plus new SWW.

Routine D (MF Broadcast)

Repetition of Inshore Waters Forecast, Gale Warnings. This (MF) routine is only made if there are gale warnings in force.

MRCC	Routine A Twice a day as shown:	Routine B Twice a day as shown:	Routine C Three hours after Routines A&B
Shetland	0710, 1910	0110, 1310	"
Aberdeen & Forth	0730, 1930	0130, 1330	"
Humber & Yarmouth	0750, 1950	0150, 1350	"
Thames & Dover	0710, 1910	0110, 1310	"
Solent & Portland	0730, 1930	0130, 1330	"
Brixham & Falmouth	0710, 1910	0110, 1310	"
Swansea, Milford Haven & Holyhead	0750, 1950	0150, 1350	"
Liverpool	0730, 1930	0130, 1330	"
Belfast	0710, 1910	0110, 1310	"
Clyde	0810, 2010	0210, 1410	"
Stornoway	0710, 1910	0110, 1310	"

1 Gale Warnings, Storm Warnings and Navigation Warnings will additionally be broadcast on receipt. Such warnings may be announced on DSC.

2 Broadcast from MRCCs Falmouth, Clyde, Brixham, Belfast and Stornoway only and occasionally Aberdeen and Forth.

3 This broadcast will only be made on MF when there are Gale Warnings in force.

Aerials & Frequencies used for broadcasting weather info for sea areas Inshore & Offshore UK

(Knowledge of the position of the aerial may be useful in improving reception)

SHETLAND

Aerials &	Saxaford, Fitful Head – 23, Lerwick 84, Collafirth
Frequencies	Wideford – 86
MF	Lerwick – 1770 kHz
Inshore Forecast/WZ	Shetland and Cape Wrath to Rattray Head
Sea Areas	Faeroes, Fair Isle, Viking, Cromarty
Additional	3 Day Fisherman's Forecast (October - March)

ABERDEEN & FORTH

Aerials &	Durness, Windy Head, Inverbervie, Craigkelly – 23,
Frequencies	Noss Head, Fifeness – 84, Rosemarkie, Gregness, St Abbs – 86
MF	Gregness – 2226 kHz
Inshore Forecast	Cape Wrath to Rattray Head and Rattray Head to Berwick
Sea Areas	Fair Isle, Cromarty, Forth, Forties, Tyne
Additional	3 Day Fisherman's Forecast (October - March)

HUMBER & YARMOUTH

Aerials &	Newton, Hartlepool, Flamborough, Guys Head,
Frequencies	Trimingham, Lowestoft – 23, Cullercoats, Ravenscar, Easington, Langham, Yarmouth – 86
MF	Flamborough – 2226 kHz
Inshore Forecast	Berwick to Whitby, Whitby to Gibraltar Point, Gibraltar Point to North Foreland
Sea Areas	Tyne, Dogger, Humber, German Bight, Thames
Additional	3 Day Fisherman's Forecast (October - March)

THAMES & DOVER

Aerials &	Walton, Southend, Bawdsey – 23, Fairlight – 84,
Frequencies	Bradwell, Langdon – 86
Inshore Forecast	Gibraltar Point to North Foreland & North Foreland to Selsey Bill
Sea Areas	Humber, Thames, Dover, Wight

SOLENT & PORTLAND

Aerials &	Boniface Down – 23, The Grove – 84, Newhaven,
Frequencies	Needles, Beer Head – 86
Inshore Forecast	North Foreland to Selsey Bill, Selsey Bill to Lyme Regis, Lyme Regis to Lands End
Sea Areas	Wight, Portland, Plymouth

BRIXHAM & FALMOUTH

Aerials &	Dartmouth, Fowey – 10, Berry Head, Lizard – 23,
Frequencies	East Prawle, Trevose, Falmouth – 84, Rame Head, Isles of Scilly – 86
MF	Lizard 2226 kHz
Inshore Forecast	Lyme Regis to Lands End, Lands End to St Davids Head
Sea Areas	Portland, Plymouth, Shannon, Fastnet, Sole
Additional	3 Day Fisherman's Forecast (October - March)

SWANSEA & MILFORD HAVEN

Aerials &	Combe Martin, St Hilary – 23, Blaenplwyf,
Frequencies	St Anns Head, Monkstone Point – 84, Hartland, Severn Bridge, Mumbles, Dinas Head – 86
Inshore Forecast	Lands End to St David's Head, St David's Head to Great Orme Head
Sea Areas	Lundy, Fastnet, Irish Sea

HOLYHEAD

Aerials & Frequencies	South Stack – 23, Great Orme – 86
Inshore Forecast	St Davids Head to Great Orme Head & Great Orme Head to Mull of Galloway
Sea Areas	Irish Sea

LIVERPOOL

Aerials &	Caldbeck, Moel y Parc – 23, Langthwaite – 84,
Frequencies	Snaefell – 86
Inshore Forecast	Great Orme Head to Mull of Galloway & Isle of Man
Sea Areas	Irish Sea

BELFAST

Aerials &	Black Mountain – 23, Limavady, Orlock – 84,
Frequencies	Slieve Martin, West Tor, Navar – 86
Inshore Forecast	Lough Foyle and Carlingford Lough, Mull of Galloway to Mull of Kintyre & Isle of Man
Sea Areas	Irish Sea, Malin

CLYDE

Aerials &	Torosay, Rhustaffnish – 10, Glengorm, South
Frequencies	Knapdale – 23, Kilchiaran, Clyde – 84, Tiree, Law Hill – 86
MF	Tiree – 1883 kHz
Inshore Forecast	Mull of Galloway to Mull of Kintyre & Mull of Kintyre to Ardnamurchan Point
Sea Areas	Rockall, Malin, Hebrides, Bailey

STORNOWAY

Aerials &	Barra – 10, Melvaig, Forsnaval, Arisaig – 23, Skriag,
Frequencies	Clettravel, Portnaguran – 84, Butt of Lewis, Drumfeam, Rodel – 86
MF	Butt of Lewis – 1743 kHz
Inshore Forecast	Ardnamurchan Point to Cape Wrath
Sea Areas	Hebrides, Bailey, Rockall, Faeroes, Fair Isle, SE Iceland, Malin
Additional	3 Day Fisherman's Forecast (October - March)

Internet

There is a wealth of useful weather information available on the Internet, including actual and forecast weather charts and hourly weather reports from buoys and lightships around Britain. This information is particularly valuable when planning a passage, and is a good way of keeping in touch with the weather when ashore. See page 31 for a selection of useful sites.

See page 14 for the Marine Observation Sites in the vicinity of the UK, providing the opportunity to dial up a site you may be planning to visit.

Weather information and forecasts available in the UK

The most popular and widely used forecasts for sea areas of northwest Europe are the Shipping Forecasts broadcast by the BBC on Radio 4. French, German and Spanish fishermen swear by them. The script is written in the Met Office's National Meteorological Centre. All forecasts broadcast by the Coastguard also originate from the NMC, as do all gale warnings for UK sea areas. Many of the following basic services are available free within the context of the requirements of the Safety of Life at Sea (SOLAS) Convention and are provided by the Met Office in association with the Maritime and Coastguard Agency (MCA). However, the Met Office also offers a range of additional services via telephone, fax and the internet providing, for a modest fee, more detailed and site specific information.

SHIPPING FORECASTS

Shipping forecasts are broadcast on Radio 4 Long Wave (1515 metres, 198kHz) at 0048, 0536, 1201 and 1754 daily; and on Radio 4 VHF at 0048, 0536 and 1754 daily and 1201 at weekends. These are clock times.

Gale warnings are part of the shipping forecast service and are broadcast at the earliest juncture in the Radio 4 programme after receipt from the Met Office, and also after the next following news bulletin.

Each Shipping Forecast Bulletin comprises several parts:

General synopsis

This is very important because it gives the information you need to estimate the time of occurrence of any changes which are forecast for individual sea areas. It tells you about depressions, anticyclones, troughs and fronts, which will control the winds and weather over the sea areas in the forecast period. It tells you where they were just before the broadcast was written, which way they are moving and where they are expected to be at the end of the forecast period. Careful interpolation on the basis of the General Synopsis will often double the amount of information you can get from the sea area forecast which follows. The meanings of terms which are used are given on page 20-21.

Forecasts for sea areas

For each sea area or group of areas around the British Isles, a forecast is given of wind, weather and visibility for the following 24 hours. The areas are given in fixed sequence and to save time the words wind, force, weather, visibility are omitted. The meanings of the terms used are given on pages 15 through 19.

Reports from coastal stations

The 0048 and 0536 bulletins end with a list of weather reports from coastal stations around the British Isles. These give the reported wind direction and force; significant weather, if any (if the weather is fair or fine nothing is said); visibility in miles or metres; barometric pressure in millibars and the barometric tendency (i.e., whether rising, falling or steady). These reports are very useful because, not only do they tell you what the weather is nearest to you, but they give you enough detail with the General Synopsis to enable you to construct your own weather map. Do not forget though, if you use them to draw a weather map, the coastal reports are always for a later time than the 'main chart time' (0001, 0600, 1200, 1800 GMT) for which the positions of depressions, troughs etc. are given in the General Synopsis. If you are constructing your own weather map, and you should do it for the time of the coastal station reports, you will need to move on the troughs, and fronts, etc., and adjust the positions given in the General Synopsis to the time of the coastal reports, using the speeds given in the General Synopsis.

Increasingly, manned weather reporting stations are being replaced by automatic ones, and it is important to be aware of their limitations. Automatic stations are not yet capable of reporting 'weather', i.e. whether it is raining, snowing etc, or of measuring visibility in excess of five miles. So if the visibility is reported as five miles, it does not necessarily mean that it is only five miles. Even more importantly, if the visibility is reported as, say, 500 metres, you have to judge whether the low value is due to fog, rain or snow.

Inshore waters forecasts

Forecasts for inshore waters (up to 12 miles offshore) around Britain are broadcast at 0053 and 0540 clock time on BBC Radio 4 (198kHz, 1515m) and VHF, and include at 0053 the most recent available actual weather reports from a selection of coastal stations.

Land area forecasts

These are important because:

a) They include an outlook beyond the period of the detailed forecast and the shipping forecast. On a fairly recent occasion, stormy weather was predicted in the weekend outlook at the end of a Friday lunchtime forecast. This was not mentioned in the shipping forecast since the change was not due until well after the end of the period covered by the shipping forecast, but sailors who took action on the land area forecast saved themselves a lot of trouble and inconvenience, while some who did not heed the forecasts were shipwrecked and taken ashore by the rescue services.

b) They may give details about coastal weather for which there is not space in the shipping bulletin.

FORECASTS AND REPORTS ON LOCAL RADIO

All local radio stations near the coast broadcast weather forecasts for the local area and many of them include information on winds and weather over the nearby coastal waters, particularly during the summer months. Since the time and content of the various broadcasts are frequently adjusted, the detailed information is not included in this book. However, you will find that the best times to tune into your local radio station for a weather bulletin are between 30 and 35 minutes past 0600, 0700 and 0800 and/or sometime between 0650 and 0715, and again between 0750 and 0815, and between 1700 and 1830.

FORECASTS VIA THE COASTGUARD

Coastguard Marine Rescue Co-ordination Centres (MRCCs) and Sub-Centres (MRSCs) provide a comprehensive schedule which includes the broadcast of Gale Warnings, Strong Wind Warnings, those parts of the Shipping Forecast relevant to the area served by the particular CG station, and an Inshore Waters forecast.

GALE & STRONG WIND WARNINGS

Gale warnings form an important part of the weather services provided by national and coastal radio stations and via Navtex. Always remember however that warnings are issued mostly for winds of gale force (34 knots) and above, because at this strength, the wind becomes a hazard for commercial shipping and trawling. The 'yachtsman's gale' is more like Force 6 (22 to 27 knots) while for many small sailing and motor boats Force 4 (11 to 16 knots) is the limit of safety.

A strong wind warning service operates for the benefit of small craft in the coastal waters of Britain. These warnings are disseminated via the Coastguard MRCCs and MRSCs (on receipt and at four-hourly intervals thereafter). Many BBC and independent local radio stations on or near the coast also broadcast these warnings.

Warnings are issued whenever winds of Force 6 or above are expected over the coastal strip (waters up to 5 miles offshore). The warnings cover the period up to 12 hours ahead and, whenever possible, advance warning of up to six hours is given. A warning is also issued if there is a Gale Warning in force for the sea area even though the winds are not expected to reach Force 6 in the coastal strip. The warning will then state that winds are not expected to reach Force 6 in the coastal strip, but that there is a warning of gales to seaward. Note that strong wind warnings automatically expire 12 hours after issue.

FORECAST FOR FISHING FLEETS

Three-day forecasts, primarily for the benefit of the fishing fleets in the North Sea and South Western Approaches, are broadcast by Aberdeen, Humber and Falmouth Coastguard stations from 1 October until 31 March. An initial announcement is made on 2182kHz and the broadcast is on 2226kHz. The forecasts for Viking, Cromarty, Forth, Forties, Fisher and Fair Isle are broadcast by Aberdeen at 2020 and repeated the next morning at 0820. Forecasts for Tyne, Dogger, German Bight, Humber and Thames, by Humber at 2110 and repeated the next morning at 0910. Forecasts for Plymouth, Fastnet, Shannon, Sole, Fiitzroy by Falmouth at 2150 and repeated the next morning at 0950.

TELEVISION AND TELETEXT
WEATHER FORECASTS

Most BBC and ITV channels give regular coverage around news times and near close down. The chart shown is usually the latest actual chart, but in some of the longer presentations, a forecast chart is also included. Both Ceefax and Oracle services include weather information and forecasts. Go to Ceefax page 409 or Teletext page 108 for an inshore waters forecast.

SPECIAL FORECAST SERVICES

'Talk to a forecaster' is an interactive fax or phone service from the Met Office accessible from anywhere in the world, and includes the opportunity to consult a forecaster or receive fax forecasts abroad. Payment is online by credit card - see page 39.

Clubs may, for a fee, make arrangements for the provision of special forecasts and warnings for particular sailing events. Marinecall fax can be a very useful part of such a service.

The Met Office operates a customer centre 24 hours a day, 7 days a week for all enquiries. The telephone number is 0870 900 0100 and the fax number is 0870 900 5050. For long offshore passages weather routeing services are available (see page 39).

MARINECALL

Marine weather services by telephone and facsimile provide a high quality service which is justifiably the envy of sailors in other countries. **Marinecall** offers 10 day inshore waters forecasts, offshore waters forecasts and European sailing forecasts. The 10 day inshore waters forecast starts with a 48 hour inshore waters forecast for the coastal area and up to 12 miles offshore. This is followed by a 5 day sea area forecast and accompanied by a national forecast for days 6-10 and outlook for the month ahead.

The **Marinecall fax** service provides hard copy versions of what is available by telephone along with the all important supporting weather maps, both actual and forecast. There is a choice from four options. For each coastal zone you can select either a *standard* or *advanced* forecast for either inshore or offshore waters, with the following specifications:

Marinecall - standard inshore coastal: a two page bulletin comprising:

- A 48 hour forecast.
- Weather maps for today and tomorrow.

Marinecall - standard offshore: comprising:

- A two to five day planning forecast.
- Weather maps.

Marinecall - advance inshore coastal: a 3-page bulletin comprising:

- The general weather situation, details of any gale or strong wind warnings, a forecast for the next 48 hours of wind direction and speed including probability and strength of gusts, weather, visibility and sea state.
- A coastal location forecast in tabular format plotting the changing weather picture for the current hour and the next five hours for four key sailing points within each coastal area.
- A weather map for today and tomorrow.

Marinecall - advance offshore: a 3-page bulletin comprising:

- A two to five day planning forecast.
- weather maps in 4-up format.
- A significant wave height contour graph.

Details of coastal and offshore areas, telephone numbers and costs are on pages 34 to 36.

For regular users of Marinecall fax a subscription service is available - Marinecall fax direct - offering considerable savings. Also RYA members qualify for a 10% discount on all Marinecall fax services.

MARINECALL MOBILE

SMS (Short Message Service)

This is limited to a maximum of 160 characters to convey a current weather report plus a 6-hour forecast of temperature, wind speed and direction, visibility and percentage chance of precipitation.

Text MC (or MC SUB if you want a daily forecast on subscription) plus name of coastal location (see pages 37-38), and send to 83141.

Internet

Marinecall forecasts can be purchased either individually or on monthly subscription from: **www.marinecall.co.uk**

The language of weather bulletins

WIND DIRECTION

The wind direction is always the direction **from** which the wind is blowing. **Veer** means a clockwise change in wind direction, e.g. **from** west to northwest. **Back** means an anticlockwise change in direction, e.g. from northwest to west.

WIND STRENGTH

In sea area forecasts wind strengths are mostly given in terms of the Beaufort Force - see table on pages 16 and 17.

GUSTS

The Beaufort Scale of wind force categorises the mean or average wind. For winds of gale force and above, the strength of gusts is also included at a value approximately 25% stronger than the mean wind. However, gusts must be expected at all wind speeds especially in cold unstable airstreams when 25% over the mean must be regarded as the norm.

WARNINGS

Warnings are issued for:

Gale

If the mean wind is expected to increase to Force 8 (34 knots) or over, or gusts of 43 knots or over are expected. Gusts as high as 43 knots may occur with the mean wind below 34 knots in cold, unstable and showery airstreams.

Severe gale

If the mean wind is expected to increase to Force 9 (41 knots) or over, or gusts of 52 knots or over are expected.

Storm

If the mean wind is expected to increase to Force 10 (48 knots) or over, or gusts of 61 knots or over are expected.

Winds above Force 10 can only be of academic interest to yachtsmen and will not be detailed here.

The words *imminent, soon* and *later* have precise meanings as follows:

Imminent – within 6 hours of issue of the warning.
Soon – 6 to 12 hours from time of issue.
Later – beyond 12 hours from time of issue.

BEAUFORT SCALE OF WIND FORCE

Beaufort number	General Description	Sea Criterion	Landsman Criterion	Limits of velocity in knots
0	Calm	Sea like a mirror	Calm; smoke rises vertically.	Less than 1
1	Light air	Ripples with the appearance of scales are formed, but without foam crests.	Direction of wind shown by smoke drift but not by wind vanes.	1 to 3
2	Light breeze	Small wavelets, still short but more pronounced. Crests have a glassy appearance and do not break.	Wind felt on face; leaves rustle; ordinary vane moved by wind.	4 to 6
3	Gentle breeze	Large wavelets. Crests begin to break. Foam of glassy appearance. Perhaps scattered white horses.	Leaves and small twigs in constant motion. Wind extends light flags.	7 to 10
4	Moderate breeze	Small waves becoming longer; fairly frequent white horses.	Raises dust and loose paper; small branches are moved.	11 to 16
5	Fresh breeze	Moderate waves, taking more pronounced long form; many white horses are formed. Chance of some spray.	Small trees in leaf begin to sway. Crested wavelets form on inland waters.	17 to 21
6	Strong breeze	Large waves begin to form; the white foam crests are more extensive everywhere. Probably some spray.	Large branches in motion; whistling heard in telegraph wires, umbrellas used with difficulty.	22 to 27
7	Near gale	Sea heaps up and white foam from breaking waves begins to be blown in streaks along the direction of the wind.	Whole trees in motion; inconvenience felt when walking against wind.	28 to 33
8	Gale	Moderately high waves of greater length; edges of crests begin to break into spindrift. The foam is blown in well-marked streaks along the direction of the wind.	Breaks twigs off trees; generally impedes progress.	34 to 40

9	Severe gale	High waves. Dense streaks of foam along the direction of the wind. Crests of waves begin to topple, tumble and roll over. Spray may affect visibility.	Slight structural damage occurs (chimney-pots and slates removed).	41 to 47
10	Storm	Very high waves with long overhanging crests. The resulting foam in great patches is blown in dense white streaks along the direction of the wind. On the whole the surface takes on a white appearance. The tumbling of the sea becomes very heavy and shock-like. Visibility affected.	Seldom experienced inland; trees uprooted; considerable structural damage occurs.	48 to 55
11	Violent storm	Exceptionally high waves. The sea is completely covered with long white patches of foam lying along the direction of the wind. Everywhere, the edges of the wave crests are blown into froth. Visibility affected.		56 to 63
12	Hurricane	Air filled with foam and spray. Sea completely white with driving spray. Visibility very seriously affected.		Greater than 63

In land area forecasts winds are always given in terms of moderate, fresh, etc. which are defined as follows:

Beaufort number

Calm	0
Light	1-3
Moderate	4
Fresh	5
Strong	6-7
Gale	8

Brisk is occasionally used in place of Fresh

FRENCH BEAUFORT SCALE OF WIND FORCE

Degrés	Termes	Vitesse descriptifs	Vitesse moyenne	Ètat de la mer moyenne en noeuds en km/h
0	Calme	<1	<1	Comme un miroir
1	Très légère brise	1 - 3	1 - 5	Quelques rides
2	Légère brise	4 - 6	6 - 11	Vaguelettes ne défereant pas
3	Petite brise	7 - 10	12 - 19	Les moutons apparaissent
4	Jolie brise	11 - 16	20 - 28	Petites vagues, nombreux moutons
5	Bonne brise	17 - 21	29 - 38	Vagues modérées, moutons, embruns
6	Vent frais	22 - 27	39 - 49	Lames, crêtes d'écume blanche, embruns
7	Grand frais	28 - 33	50 - 61	Lames défereantes, trâinées d'écume
8	Coup de vent	34 - 40	62 - 74	Tourbillions d'écume à la crête des lames, traînées d'écume
9	Fort coup de vent	41 - 47	75 - 88	Lames défereantes, grosses à énormes, visibilité réduite par les embruns
10	Tempête	48 - 55	89 - 102	Lames défereantes, grosses à énormes, visibilité réduite par les embruns
11	Violente tempête	56 - 63	103 - 117	Lames défereantes, grosses à énormes, visibilité réduite par les embruns
12	Ouragan	>64	>118	Lames défereantes, grosses à énormes, et plus visibilité réduite par les embruns

Les vitesses se rapportent au vent moyen et non aux rafales.

VISIBILITY

In sea area forecasts, visibility descriptions have the following meanings:

Good	More than 5 nautical miles
Moderate	2 to 5 nautical miles
Poor	1,000 metres to 2 nautical miles
Very Poor	Less than 1,000 metres

In land area reports and forecasts fog is defined as:

Fog	Visibility 200 to 1,000 metres
Thick Fog	Visibility less than 200 metres
Dense Fog	Visibility less than 50 metres

In coastal station reports and aviation forecasts the definitions are:

Mist or Haze	Visibility 1,000 to 2,000 metres
Fog	Visibility less than 1,000 metres

WEATHER

This is not included in reports from automatic stations.

The terms rain, snow, hail etc are obvious enough but the use of the word **fair** calls for some definition. The weather is described as fair when there is nothing significant i.e. no rain, fog, showers, etc. It may or may not be cloudy.

PRESSURE AND PRESSURE TENDENCY

The General Synopsis often gives the values of the pressure at the centres of important weather systems, while the coastal station reports give recorded atmospheric pressure at a selection of stations and the pressure tendency. The millibar is the unit used for pressure in shipping bulletins and on most charts published in the press and elsewhere. Some countries have replaced millibar by hectopascal as the name for the standard unit of pressure. There is no numerical difference between the two.

One millibar (or mb) = one hectopascal (or hPa)

The terms used for pressure tendency in the coastal station reports are defined as follows:

Steady	Change less than 0.1mb in 3 hours
Rising slowly or falling slowly	Change 0.1 to 1.5mb in last 3 hours
Rising or falling	Change 1.6 to 3.5mb in last 3 hours
Rising quickly or falling quickly	Change 3.6 to 6.0mb in last 3 hours
Rising or falling very rapidly	Change of more than 6.0mb in last 3 hours
Now falling, now rising	Change from rising to falling or vice versa within last 3 hours

BEWARE of reading too much into reports of rising slowly, falling slowly, now falling and now rising if general pressure changes are small. Every day there are small ups and downs in pressure all over the world due to the atmospheric tide. In the south of the UK the tidal pressure variation is just under 1mb. At the equator it is 3mb. The highest values of pressure due to this tide are at 1000 and 2200, the lowest at 0400 and 1600: the same local times everywhere in the world. So if at 0400 and 1600 the pressure is reported as falling slowly it does not mean the weather is likely to or beginning to deteriorate. Similarly if at 1000 and 2200 the pressure is reported as rising slowly it says nothing about improvement.

THE GENERAL SYNOPSIS

Depression

A **depression** is synonymous with a **low** (i.e. low pressure system with a central vortex) or a cyclone (but only the relatively weak tropical cyclones are called depressions). A depression is a cyclonic vortex in the atmosphere in which the winds circulate anticlockwise in the northern hemisphere (clockwise in the southern hemisphere), and blow slightly inwards towards the centre. Depressions in middle latitudes vary enormously in size and energy. Their diameter may be anything from 100 to 2,000 miles with winds of from about 10 to over 70 knots at the surface and central pressure from below 950 millibars in a really deep depression to perhaps as high as 1,025 millibars in a very shallow one. Their speed of movement may be anything up to 60 knots. In a newly formed depression the circular spiralling motion of the air may extend upwards to only a thousand metres or so at most, while in an old depression it may extend upwards to over 15,000 metres.

A **deepening** depression is one in which the central pressure is falling, and in which the winds and rain must be expected to increase. In a **filling** depression the reverse applies. A **vigorous** depression may be large or small but is characterised by strong winds and a lot of rain. A **complex** depression has more than one centre of low pressure.

The simplest form of depression is a perfectly circular system in which the winds are the same speed all the way round and decrease gradually as you move out from the centre. The nearest approach to this ideal is found in the tropics, but in middle latitude, their shape and wind distribution vary greatly. In some, the strongest winds are near the centre, in others the strongest winds may be 500 miles out from the centre.

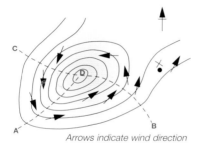

Arrows indicate wind direction

Trough of low pressure

In most depressions the cloud and rain tend to be concentrated in bands extending outwards from somewhere near the centre. These bands of weather may be from 10 to 200 miles wide and are called **troughs of low pressure** because the pressure is lower along them than at other points the same distance from the centre of the depression. There is a very definite relation between pressure and wind; the wind blows according to the gradient of pressure. The greater the fall of pressure from one point to the next the stronger the wind between the two points. So with a trough of low pressure; the more marked it is, the stronger the winds associated with it and the greater the change in the wind direction from one side of it to the other.

The simple diagram above will help to make this point clearer.

D marks the centre of a depression. The line DA marks the axis of a **vigorous** trough of low pressure in which there is a change of wind of over 90 degrees as you go from one side to the other, you would expect a lot of rain mainly on the forward side of it. DB marks a less vigorous trough, but still with quite a marked wind change and probably a fair amount of cloud and rain. DC is a relatively **weak** trough with comparatively little wind change and probably only occasional rain or a belt of showers.

Depressions can move in any direction but most frequently from west to east. Whether the depression is moving or not, its troughs will usually be circulating round its centre. If you are at X and the trough moves across you, you can see from the diagram that the winds will back and increase as the trough approaches and veer as it passes. The passage of even the weakest trough may make a lot of difference to your attempt to round a windward mark. If you are at X and the depression moves ENE to the north of you, the wind will veer from southerly ahead of the low to northerly behind it. The veer is unlikely to be steady and will in fact be concentrated in the passage of the troughs, while ahead of the troughs you may experience a temporary backing of the wind. This is typical of what is met with in practice, but there are an awful lot of variations on the theme. Finally, if you are at X and the depression moves ESE keeping its centre south of you, the winds will back from southerly to northerly. Any troughs will again be an added complication.

Troughs may fill or deepen independently of the parent depression, and sometimes a deepening trough will develop its own circulation and a new low is formed – a **secondary depression**.

Front

There is little that needs to be said about fronts because they are simply a particular type of trough of low pressure, and troughs have been discussed in the previous paragraph. A **front** is, in fact, a trough of low pressure in which there is a change in air mass from one side to the other. In a **warm front** the air mass changes from cold to warm as the warm frontal trough passes, and in a **cold front** the air mass changes from warm to cold as it passes. The troughs at DB and DA are typical positions for warm and cold fronts respectively in the circulation of a depression, with the warm air in between them. Troughs in which there is an air mass change are usually more vigorous than those in which there is none. In fact wherever you find an air mass change you will usually find some sort of trough of low pressure. Fronts have a number of typical weather characteristics which are discussed in most books on the weather. Their wind change characteristics are the same as for the trough. Incidentally, in land area forecasts, fronts are rarely mentioned and are usually referred to as troughs. The same sometimes applies in the shipping forecasts.

Anticyclone

The word anticyclone obviously means something contrasting with cyclone. An anticyclone or high has a **high** central pressure relative to its surroundings, fair or fine weather and light winds circulating clockwise around the centre in the northern hemisphere and anti-clockwise in the southern, and blowing slightly outwards. They vary in size from perhaps 200 miles across for the small high that accidentally finds itself between two lows, to some 2,500 miles across for some of the large and very persistent anticyclones which are responsible for the longer spells of dry weather. An anticyclone is said to **build** if its pressure is rising and to **decline** or **weaken** if its pressure is falling. If the pressure is falling very quickly it is said to be **collapsing**. Large anticyclones are usually slow-moving but some of the smaller ones which occur between depressions move quickly – perhaps as fast as 30 to 40 knots. More often, between depressions, instead of a high, with a closed circulation, one finds a **ridge of high pressure**. This is analogous to a mountain ridge between two hollows. Similarly a weak ridge may well occur between two troughs (or valleys) even in the circulation of a depression. The weather in a ridge is similar to the weather in an anticyclone though it may not last as long. In a weak ridge between two fast moving troughs it may be little more than a bright period.

As you move outwards from the centre of an anticyclone or away from the axis of a ridge, the winds usually increase. If a ridge is crossing your course, the winds will decrease as it approaches, back as it passes and then increase and continue to back depending on what is coming next.

Speed

The following terms are used in bulletins to describe the speed of movement of pressure systems:

Slowly	up to 15 knots
Steadily	15 – 25 knots
Rather quickly	25 – 35 knots
Rapidly	35 – 45 knots
Very rapidly	over 45 knots

STANDARD DEFINITIONS OF STATE OF SEA

Sea state	Height of waves in metres	Definition English	French
0	0	calm – glassy	calme – plate
1	0 to 0.1	calm – rippled	calme – ridée
2	0.1 to 0.5	smooth	belle
3	0.5 to 1.25	slight	peu agitée
4	1.25 to 2.5	moderate	agitée
5	2.5 to 4	rough	forte
6	4 to 6	very rough	très forte
7	6 to 9	high	grosse
8	9 to 14	very high	très grosse
9	over 14	phenomenal	énorme

Writing down the shipping bulletin

Shipping bulletins are broadcast at normal reading speed. To write it down you will need a prepared form with at least all the sea areas listed and some form of shorthand notation for the majority of the standard words and phrases. Forms suitable for recording shipping forecasts are available from the RYA and the Royal Meteorological Society, Reading RG1 7LL. To download them from the RYA website, enter the URL www.rya.org.uk/KnowledgeBase/weather then click on Metmaps under Related Links. Page 2 is the form for entering directly from the broadcast. There is nothing to stop you devising your own shorthand so long as you can remember afterwards the meaning of what you have written, but there is a lot to be said for using a notation which has been evolved by those with considerable experience in using shipping bulletins and which includes a number of standard international weather map symbols. Once you are familiar with the more common of these international symbols you will be able to appreciate at a glance the information on any weather map which you may see displayed in clubs or at ports of call.

THE GENERAL SYNOPSIS

Few international symbols are involved here except of course for the points of the compass – N, S, NW, SW, etc – but it is not possible to write down the synopsis until you have practised a simple shorthand using initial letters for terms such as depression, anticyclone, low, high, warm front, cold front, occlusion, trough, ridge etc., and for the various sea areas which are usually referred to in giving the positions of the weather systems. One very useful hint is to use an oblique stroke to denote the passage of time, and this applies particularly to the sea area forecasts as we shall see. Also, it is often a good idea to denote movement by an arrow.

THE SEA AREA FORECAST

Wind – The simplest forecast is in the form 'northwest 5', which is obviously abbreviated as 'NW 5'. Often however, we have something like 'northwest 4 to 5 at first, backing southwest and increasing to 7 to gale 8 by end of the period' which should be written as 'NW 4–5/SW 7–8', and all the rest of the words can be inferred from this simple shorthand form. Note particularly the use of an oblique stroke to distinguish between 'at first' and 'later'. 'In the South at first' abbreviates to 'In S/' whereas 'In the South later' would abbreviate to '/in S'. The words 'temporarily', 'occasionally', 'locally' are often used and should be abbreviated to 'tem', 'occ', 'loc'.

Weather – This is always given in terms such as 'fair', 'showers', 'rain' etc. Here international shorthand should be used and you can choose between the Beaufort letter notation or the international weather map symbols as given on page 24. There is something to be said for using the international map symbols because you can then plot these directly onto your map, but the former are much easier to learn and you can readily turn them into plotting symbols at your leisure after the broadcast. The phrases 'at first' and 'later' are often used and, again, an oblique stroke comes in very useful. For instance 'rain at first, showers later' can be abbreviated to 'r/p'. This sort of detail should always be taken down as it almost certainly ties in with a change of wind and the passage of an important weather system through the sea area.

Abridged Beaufort weather notation and international plotting symbols

Beaufort letter		Plotting symbol
r	rain	●
d	drizzle	୨
s	snow	✳
p	shower	▽
h	hail	△
th	thunderstorm	⏃
q	squall	⋎
m	mist	=
f	fog	≡
z	haze	∞

For heavy precipitation, capital letters are used e.g. R – heavy rain.

It is useful to know that these plotting symbols are used operationally, and 'plotted charts' (like the one below) are available on www.wetterzentrale.de.

Visibility – A straightforward abbreviation of 'g' for 'good', 'm' for 'moderate', 'p' for 'poor' is all that is required here, remembering again to use a vertical stroke to denote a passage of time and also to take down all the details which are given about fog. Fog will be discussed in more detail later under the discussion on weather hazards beginning on page 29.

COASTAL STATION REPORTS

As with the sea area forecasts you need a prepared form with the coastal stations already listed, and columns for the reports which are always given for each station in the sequence; wind, significant weather (fair and fine are not 'significant' and are not mentioned – if nothing is said about the weather you should assume that it is fair or fine), visibility in miles or metres, barometric pressure in millibars, and finally pressure tendency (i.e. whether the barometer is rising or falling, and how rapidly, or whether it is steady). The same shorthand should be used as for the sea area forecasts.

There is no need to write down the words miles or metres as one or two figures will always be miles and three or four figures will always be metres. The pressure tendency should be abbreviated to s for steady, r for rising, rs for rising slowly, etc: or you can use a stroke inclined at various angles according to the way you would observe the pressure tendency on a barograph. The definitions of the terms used for pressure tendency are given on page 19.

Using the shipping forecast and other weather bulletins

Write down at least the General Synopsis, the forecast for the sea areas in which you will be sailing, the forecast for adjacent sea areas, the reports from coastal stations nearest to your area, and the outlook from the land area forecast. If you are sailing close inshore write down also the regional land area forecast particularly details about the wind.

Having got all this information you can use it to decide the best strategy and tactics. Whether you are ocean racing or merely crossing Lyme Bay, if you set off on the wrong tack for an expected change in wind it can cost you many hours.

If you are under power and fail to appreciate that a wind change is going to coincide with an adverse tide it can give you a most uncomfortable and perhaps dangerous short steep sea, or make it impossible to cross the bar when you were intending to make port before dark.

Many such situations can be avoided by careful attention to the weather forecasts. The weather can change very quickly. Forecasts are on the air at least every six hours and you should listen to all of them. If your radio goes wrong put into harbour for repairs.

The following example will show you the sort of detail which you can glean from careful attention to the shipping bulletin.

The morning Shipping Forecast included the following:

General Synopsis – 'a trough of low pressure will move eastwards across the British Isles and is expected to lie from Viking to German Bight at midnight tonight'.

Sea Area Forecast – 'Humber, Thames, Dover – southwest veering northwest, 4 or 5; rain or drizzle followed by showers; moderate or poor with fog patches, becoming good'.

'Wight, Portland – southwest veering northwest, 3 to 4; rain or drizzle then showers; visibility moderate or poor with fog patches, becoming good'.

'Plymouth – northwest, 3 or 4; showers becoming fair; good'.

Let us assume that you are hoping to sail in sea area Dover. The wind force given for the group of areas which includes Dover is 4 or 5 but since the next group to Dover is given as 3 or 4 we can infer that the wind in Dover is likely to be at the Force 4 end of the range.

The southwesterly wind is obviously associated with the rain, drizzle and fog patches ahead of the trough of low pressure, and the northwesterly wind with the showers and good visibility behind the trough. The trough has all the characteristics of a cold front. The forecast of northwesterly winds for Plymouth tells us that the trough is already past that area and we can infer that it is moving over the western edge of Portland. We are told that it will be over German Bight by midnight so we can interpolate its movement between these two sea areas and get a reasonably good forecast of its time of crossing sea area Dover. Hence we can derive our own forecast of the time of veer of wind in Dover and the time of clearance of the rain, drizzle and fog. We can in fact plan a fair passage across the Channel from late morning onwards with a following wind, good visibility and just a few showers. The coastal stations reports from south coast stations would be found to confirm the inference as to the present position of the trough.

The more you practise using the shipping forecasts the more interesting and useful you will find the bulletins and reports from all sources, even those for the more remote areas such as southeast Iceland. Each item of information becomes a piece in a jig-saw which you need to complete the whole picture.

But what if your sources of information disagree or are inconsistent? Suppose Jersey Radio gives a different forecast from Niton, or Scheveningen tells a different story from North

Foreland. It is no use denying that this happens. It does. But do not despair. Remember two things. First, that both forecasts must be based on the same initial data – they cannot be contrary in that sense; second that you already have, or should have, a good idea as to what the weather chart looks like in terms of depressions and anticyclones, troughs and ridges. This cannot be in dispute. So you, with your weather map in front of you, are in a good position to sort out for yourself what is the best possible forecast for your own particular area. You can even try to identify the reasons for the two divergent forecasts in terms of what is happening on the chart. And what is even more important, the lapse of time since the forecast was prepared means that you have additional information in terms of the weather you have observed and are observing which will help you to sort out the answer. What is the barometer doing, the wind, the cloud, and the sea? Even the smallest amount of information is all part of the overall pattern of movement and change in the atmosphere. And if you can go even a small way to understanding it, it will provide a new dimension to life afloat.

READING WIND SPEED AND DIRECTION FROM WEATHER MAPS

Many weather maps have a scale of wind speed in one corner called the 'geostrophic scale' (see illustration below). Using a pair of dividers, take the distance between adjacent isobars on the weather map over the area of interest and read from the scale the wind speed for the latitude of the area. This is the actual or predicted wind speed at a height of about 500 metres. The wind on the water is some 10% to 20% less than this. If you are sailing in different latitudes it is important to appreciate that for a given isobar spacing (pressure gradient) the wind is much stronger in low latitudes compared to high latitudes.

GEOSTROPHIC WIND SCALE
IN KNOTS FOR 4 MB INTERVALS

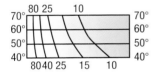

SCALE OF NAUTICAL MILES

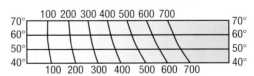

The reading of the wind from the geostrophic scale for a given latitude and pressure gradient is always the same for the same map projection and scale, so it is worth cutting one out. Using, for instance the scales above, measure the distance apart of isobars at 4mb intervals in miles on any weather map; set your dividers to this distance on the scale on the right. Transfer this setting to the scale on the left for the appropriate latitude and read off the wind speed. Note: Some countries use isobar spacing of 5 millibars.

No scale is needed to read the wind direction from a weather map. The wind on the water is the direction given by the run of the isobars less about 15 degrees.

Coastal winds

Winds within about 10 miles of the coast are influenced by the contours of the nearby land, by the local generation of sea and land breezes, by the state of the tide, and by the fact that air blowing over water is subject to a different frictional force from air blowing over land. These coastal influences may cause a difference of as much as 10 to 15 knots in the wind at points only a few kilometres apart, even when the coast is fairly flat. Near mountains a local increase of 20-30 knots is not uncommon.

SEA AND LAND BREEZES

Sea breezes develop when the land becomes warmer than the sea, and the direction of the pressure gradient wind is from land to sea. Typically, on a bright or sunny day an offshore wind at breakfast time drops to calm near the coast, and is followed by an onshore sea breeze which increases steadily, veers, extends seawards, and may reach Force 6 close to the coast by mid-afternoon.

When the gradient wind is onshore and the day is bright or sunny, the wind will either increase or decrease a few knots by afternoon depending on whether the land is to the left or right of the wind direction looking downwind.

Land breezes are experienced at night. They are strongest under clear skies and at the mouths of valleys.

FRICTIONAL EFFECTS

Land, particularly where there are trees and buildings, exerts a drag on the air. The drag over the water is much less. These frictional forces not only slow the air down but also cause a change of direction, and the direction over the water is about 15° different from the direction over the land. Winds blowing nearly parallel to the coast will converge or diverge depending on whether the greater friction is to the right or left of the wind direction. With convergence (east winds on a south facing coast, west winds on a north facing coast, etc) a band of stronger winds is experienced within about three miles of the coast; up to 10 knots stronger in some cases. Conversely with divergence (west winds on a south facing coast, etc) winds may be that much lighter near the coast, except when land or sea breezes are blowing.

TIDAL EFFECTS

A change in tide may influence the wind due to:

- a change in friction as the height and length of the wave changes
- a change in water temperature, particularly near an estuary
- a change in temperature of sandbanks or mud flats as they emerge from or disappear beneath the water

More detailed information will be found in: *RYA Weather Handbook* by Chris Tibbs published by the RYA.

Weather hazards

GALES

Gales which are due to depressions do not spring up without warning. All inshore sailors can avoid them and so too can many offshore sailors. Squalls and thunderstorms are a different matter and will be discussed later.

How does one receive warning of an approaching gale? The most obvious answer is to keep a listening watch on Radio 4 or the nearest coast radio station.

Your barometer or barograph will also give you good warning of an approaching gale. A fall of pressure of over 8 millibars in 3 hours is almost certain to be followed by a gale whatever your wind is to start with, and a fall of pressure of over 5 millibars in 3 hours is almost certain to be followed by a Force 6 (the yachtsman's gale). If your wind is Force 3 or less when you observe these tendencies your barometer will have given you about 4 to 8 hours warning. A very rapid rise in pressure after a trough has passed is also indicative of a gale and the same figures apply – a rise of over 8 millibars in 3 hours for a Force 8 and over 5 millibars in 3 hours for a Force 6. You must of course make allowances for your own movement, either towards or away from the depression. Buy's Ballots Law – if you stand with your back to the wind, low pressure is on your left hand side – will tell you which way you are going relative to the depression.

If the barometer is falling rapidly and clouds are increasing rapidly – particularly if the upper clouds are moving fast and are well veered to the surface winds – then fear the worst and do not be caught on a lee shore. Slower changes in the barometer reading do not necessarily preclude a gale, but they are less definitive.

STRONG LOCAL WINDS

If the wind is blowing almost parallel to the coast, or at an angle of up to about 25 degrees, be prepared for a local increase of wind up to 10 miles from the coast. This is especially marked on the edges of anticyclones when a local increase of over 10 knots may occur. For instance, an easterly wind blowing down the English Channel in the circulation of an anticyclone to the north, while only Force 4 over most of the Channel, may be Force 6 or even 7 along the English coast. An outstanding example is with a northeasterly over Biscay when an increase of as much as 20 knots frequently occurs off Cape Finisterre.

Another wind which needs watching is the sea breeze. On a sunny summer's day, if conditions are right, the sea breeze may enhance the actual wind to give a local increase of a good 10 knots just along the coast. A gentle breeze in a warm and sheltered harbour is often unrepresentative of conditions out at sea.

SQUALLS AND THUNDERSTORMS

The arched line of black cloud associated with a squall can usually be seen as it approaches and so it gives its own warning, but only a brief one, of about half an hour. The only thing to do is to reef and make for the lightest part of the cloud. Having weathered the squall you can usually take it that another one is unlikely for 4 to 6 hours.

The advancing dark mass of threatening cloud associated with a mature thunderstorm is distinguishable from that of an advancing depression by the lack of freshening wind and sea ahead of it. In fact, it is often heralded by a decrease in wind and an almost glassy sea. The best rule is to sail so as to leave the storm to port. By doing so, although you may not miss the associated squalls you should miss the worst of them. Your barometer or barograph will show very erratic pressure changes in a thunderstorm, jumping down and up by a millibar or two.

FOG

Two types of fog must be distinguished; land and sea fog.

Land or radiation fog

This is fog which forms over the land on a clear night. It may drift seawards from the coast but does not usually go far before dispersing, and rarely more than 2 to 3 miles from the shore. What is more, as soon as it hits the sea, it starts clearing near the surface and by the time it is 100 metres offshore it is usually clear to about mast height. So if the forecast is for fog over land clearing during the morning, you can safely go out to sea expecting the fog to be gone by the time you return to port.

Sea fog

This is one of the worst hazards at sea. It forms when warm moist air is carried by the wind over a relatively cold sea. The criterion for sea fog to form is when the dewpoint of the air is equal to or above the sea surface temperature. In winter and spring the sea is coldest inshore so fog forms more frequently along the coast than out to sea. In summer and autumn the sea is coldest away from the shore so fog forms more frequently out at sea. There are always variations in sea temperature from place to place and consequently variations in the extent and intensity of sea fog.

If the dewpoint of the air being blown across the sea is very high, and everywhere the sea temperature is lower, then **widespread fog** or **extensive fog** is forecast. If the dewpoint of the air is only a little above sea temperature and in some places may not be so, then **fog banks** are forecast. If the dewpoint of the air is only above the sea temperature in some places then **fog patches** are forecast, or, in winter and spring just coastal fog if that is appropriate.

HURRICANES

Hurricane is the name given to a tropical cyclone when it occurs in the North Atlantic or north-east Pacific. It is synonymous with a cyclone in the Indian Ocean and a typhoon in other parts of the Pacific.

STEEP WAVES

Dangerously steep waves are encountered when the wind is blowing against the current. Around Britain the danger lasts typically for 2 or 3 hours around the maximum tidal stream, given a contrary wind. A much publicised example was the 1998 Sydney-Hobart race when a new vigorous depression with 50 knot southerly winds moved over the 3 knot south going East Australian current. Considerable damage and loss of life ensued.

Sources, content, maps and tables

INTERNET – SELECTION OF WEATHER SITES FOR THE MARINER

Met Office

The Met Office has several pages of useful information for the mariner, much of which is free of charge. There are added value products available via the MetWEB pages, these pages containing similar information to that available on the Marinecall fax service. Full details can be found by following the link to MetWEB.

Met Office home page
http://www.metoffice.gov.uk

Index for Leisure marine products
http://www.metoffice.gov.uk/leisuremarine/index.html

Index for weather maps and satellite pictures
http://www.metoffice.gov.uk/weather/index.html

The Shipping Forecast from the Met Office
http://www.metoffice.gov.uk/datafiles/offshore.html

The Inshore Waters Forecast and Strong Winds from the Met Office
http://www.metoffice.gov.uk/weather/marine/inshore_forecast.html

Latest Gale Warning issued by the Met Office
http://www.metoffice.gov.uk/weather/marine/shipping_forecast.html

Listing of links to marine weather products

The site maintained by Martin Stubbs provides access to marine forecasts world-wide and other related information such as links to weather charts of interest to the mariner. There is also a section on how to obtain marine forecasts by e-mail – a not-too-expensive way of obtaining forecasts on the high seas.
http://www.users.zetnet.co.uk/tempusfugit/marine

Frank Singleton's site for small craft owners contains a wealth of information on accessing weather related sites of interest to the mariner. In particular, there are pages explaining how to access GRIB data at sea for display on a lap-top computer (GRIB data files of forecasts, for example wind forecasts, can be accessed using e-mail at sea and displayed using a suitable viewer).
http://www.franksingleton.clara.net

Selection of other links to marine related weather information

Irish coastal waters: Shipping forecast and latest gale warning via Met Eireann
http://www.met.ie/forecasts/sea-area.asp

Météo-France – marine home page
http://www.meteo.fr/meteonet_en/temps/activite/mer/cotes/cot.htm

Jersey Met – a good site for weather in and around the Channel Islands
http://www.jerseymet.gov.je

Reports from buoys (hourly), light vessels and ship reports
http://www.ndbc.noaa.gov/maps/united_kingdom.shtml

Latest actual and forecast charts via Köln University
http://www.uni-koeln.de/math-nat-fak/geomet/meteo/winfos/index-e.html

European Centre for Medium Range Weather Forecasting (ECMWF) – forecast weather charts for 3 to 6 days ahead for:
- Europe and eastern North-Atlantic
- Northern Hemisphere
- Southern Hemisphere
http://www.ecmwf.int

Satellite Imagery

Dundee University provides access to world-wide real-time imagery (NOAA and Geostationary). Note: *one must register but registration is free*
http://www.sat.dundee.ac.uk

European Space Agency for Meteosat imagery
http://www.eumetsat.int/Home/index.htm

It is important to note that sources of meteorological information on the internet are not guaranteed, and their availability must not be assumed. For safety on an offshore passage you should ensure that you are equipped at least with Navtex - see pages 7 and 41.

STATIONS WHOSE LATEST WEATHER REPORTS ARE BROADCAST IN BBC RADIO 4 SHIPPING BULLETINS

Lerwick

* Automatic stations whose reports do not include weather, i.e. if there is rain, drizzle, showers and so on.

Stornoway

Tiree *

Malin Head LH

Fife Ness

Ronaldsway

Bridlington

Valentia

Sandettie Light Vessel *

Scilly *

Greenwich Light Vessel *

Channel Light Vessel *

Jersey

STATIONS WHOSE LATEST WEATHER REPORTS ARE BROADCAST IN BBC RADIO 4 INSHORE WATERS BULLETINS

Lerwick

Stornoway

Wick *

Aberdeen

Leuchars

Machrihanish

Greenock

Boulmer

Larne

Ronaldsway

Bridlington

Valley

Liverpool (Crosby)

Gt Orme Head

Gibraltar Pt

Aberporth

Milford Haven

Sheerness

Scilly *

St Catherines Point *

Jersey

* *Automatic stations whose reports do not include weather, i.e. if there is rain, drizzle, showers and so on.*

MARINECALL Forecasts & weather reports by phone & fax
Marinecall and Marinecall fax (see page 13)

Marinecall voice inshore forecast	09068 500 4xx (key map 1)
Marinecall fax - standard inshore forecast	09060 100 4xx (key map 1)
Marinecall fax - advance 10 day forecast	09065 300 2xx (key map 1)
Marinecall voice - 2 to 5 day planning forecast	09068 500 xxx (key map 2)
Marinecall fax - standard offshore	09060 100 xxx (key map 3)
Marinecall fax - advance offshore	09065 300 xxx (key map 4)
Marinecall fax - forecast weather maps only	
today & tomorrow	09065 300 277
2 to 5 days ahead	09065 300 278
Marinecall customer services	0871 200 3985

xx - insert appropriate area number from key map 1 - page 35

xxx - insert appropriate area number from key maps 2, 3 or 4 - page 36

Telephone calls cost 60p/minute from a UK landline, fax calls cost £1.50/minute.

Membership of Marinecall Club provides access to Marinecall products in the UK. Payment can be made by credit or debit card and all RYA members are entitled to a discount. Details can be obtained from Marinecall customer services.

Marinecall Customer Centre

0871 200 3985

Email: weathercall@itouch.co.uk

Marinecall and Marinecall fax coastal area numbers

51	Cape Wrath - Rattray Head
52	Rattray Head - Berwick
53	Berwick - Whitby
54	Whitby - Gibraltar Point
55	Gibraltar Point - North Foreland
56	North Foreland - Selsey Bill
57	Selsey Bill - Lyme Regis
58	Lyme Regis - Hartland Point
59	Hartland Point - St David's Head
60	St David's Head - Great Orme Head
61	Great Orme Head - Mull of Galloway
62	Mull of Galloway - Mull of Kintyre
63	Mull of Kintyre - Ardnamurchan
64	Ardnamurchan - Cape Wrath
65	Lough Foyle - Carlingford Lough
32	Channel Islands

Coastal area boundaries - Key map 1

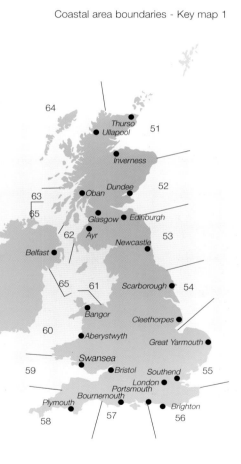

3 to 5-day Planning Forecast Areas

Dial code plus area number

Marinecall voice 09068 500 XXX

Marinecall fax - standard 09060 100 XXX

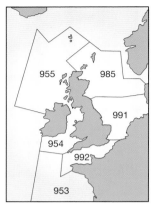

Key map 2

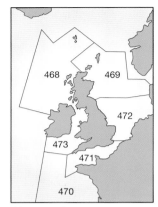

Key map 3

Marinecall fax - advance 09065 300 XXX

Key map 4

MARINECALL MOBILE (SMS)

This provides coastal weather information direct to your mobile phone. On demand at any time or by prior subscription at 0800 daily, you can receive a text forecast comprising:

- Place
- Date
- Current weather report
- Forecast for next 6 hours
- Maximum temperature
- Average windspeed (knots)
- Average wind direction (degrees)
- Visibility (km)
- Chance of precipitation (percentage)

The way in is as follows: Text MC plus name of coastal location (see pages 37-38 for list). Send to 83141.

To subscribe to a regular service text MCSUB plus name of coastal location.

COASTAL LOCATION FORECAST AREAS

The following Coastal location forecast areas are available through Marinecall Mobile.

Aberdeen
Abersoch
Aberystwyth
Aldeburgh
Allington
Amble
Anvil Point
Ardfern
Ardrossan
Bangor
Bardsey Island
Beachy Head
Beaumaris
Bembridge
Berthon Lymington Marina
Berwick
Birdham Pool
Blyth
Boston Marina
Bradwell Marina
Bridlington
Brightlingsea
Brighton Marina
Bristol
Brixham
Buckler's Hard
Burnham
Burntisland
Burrow Head
Caernarfon
Caledonia
Campbeltown
Cape Wrath
Channel Isles
Chatham
Chichester
Christchurch
Cobb's Quay
Conwy
Corpach
Cowes
Craobh
Cromer

Cuxton
Dart Marina
Darthaven
Dover
Duncansby Head
Dundee
Dungeness
Dunstaffnage
East Cowes
East Loch Tarbert
Eastbourne
Emsworth
Essex Marina
Exmouth
Eyemouth
Falmouth
Flamborough Head
Fowey Harbour
Fox's Marina
Gillingham
Glasson Dock
Gosport
Granton
Great Yarmouth
Grimsby
Guernsey
Hamble Point
Hartlepool
Haslar
Helford River
Holy Island
Holyhead
Hoo Marina
Hull
Hythe
Inverness
Iona
Ipswich
Island Harbour
Jersey
King's Lynn
Kip Marina
Kirkcudbright

Lamlash
Landguard Point
Largs Marina
Littlehampton
Liverpool
Lizard Point
London
Longships
Lossiemouth
Lowestoft
Lyme Regis
Lymington
Maryport
Mayflower Marina
Medway Bridge
Mercury Yacht Harour
Milford Dock
Montrose
Mull of Kintyre
Needles Fairway
Newhaven
Newquay
Newton Ferrers
Neyland
North Foreland
Northney Marina
Oban
Orford Ness
Padstow
Penarth
Penzance
Peterhead
Plymouth
Poole
Port Dinorwic
Port Edgar
Port Hamble
Port Medway
Port Solent
Portishead
Portland Bill
Portpatrick
Premier Swanwick
Preston
Pwllheli
Queen Anne's Battery Marina
Queenborough

Ramsgate
Rattray Head
Rhu Marina
Royal Quays Marina
Ryde
Rye Harbour
Salcombe
Salterns Marina
Selsey Bill
Shamrock Quay
Sheerness
Shoreham
Shotley Point
Solent
South Bishop
Southampton
Southend
Southsea
Southwold
Sparks Yacht Harbour
Spurn Head
St Abbs Head
St Catherines Point
St Peter's Marina
Start Point
Stonehaven
Suffolk Yacht Harbour
Sunderland
Sutton
Swansea
Tidemill Yacht Haven
Titchmarsh
Tobermory
Tollesbury
Torquay
Troon Marina
Walton-on-the-Naze
Wells-next-the-Sea
Weymouth
Whitby
Whitehaven
Whitehills
Whitstable
Woolverstone
Wootton Creek
Wyre Dock
Yarmouth Harbour

'Talk to a Forecaster'

This interactive fax or phone service includes the opportunity to speak with a Met office forecaster and is accessible from anywhere in the world. Payment is by credit or debit card over the telephone.

Full information can be obtained by dialling the Met office customer centre on 0870 900 0100.

Mediterranean Forecast Service

To consult a marine forecaster about the weather in the Mediterranean and Canary Isles phone: 08700 767 890. Calls are charged at a flat rate of £17.00.

Forecasts for Spain and the Balearics are available by fax on the same number at a charge of £3 each chart.

To obtain Mediterranean forecasts from the UK dial 09060 100 plus area code below. They are currently limited to the Balearics and the eastern coastal waters of Spain.

435	Gibraltar to Malaga		438	Valencia to Barcelona
436	Malaga to Cartagena		439	Balearic Islands
437	Cartagena to Valencia			

Weathercall

Dialling prefix 09068 500

401	Greater London		416	N W England
402	Kent, Sussex, Surrey		417	W & S Yorks & Peak District
403	Dorset, Hants, Isle of Wight		418	N E England
404	Devon & Cornwall		419	Cumbria, incl Lake District, Isle of Man
405	Wilts, Gloucs, Avon & Somerset		420	Dumfries and Galloway
406	Berks, Bucks & Oxon		421	Central Scotland & Strathclyde
407	Beds, Herts & Essex		422	Fife, Lothian & Borders
408	Norfolk, Suffolk & Cambs		423	Tayside
409	South Wales		424	Grampian & E Highlands
410	Shrops, Hereford & Worcester		425	W Highlands & Islands
411	Central Midlands		426	Caithness, Sutherland, Orkney & Shetland
412	East Midlands		427	N Ireland
413	Lincs & Humberside		430	National 3 to 5 day outlook
414	Dyfed & Powys			
415	Gwynedd & Clwyd			

Calls cost 60p per minute from a UK landline

WEATHER ROUTEING

Specialist marine weather routeing services are available from:-

MetWorks Limited

MetWorks Ltd provides a full weather routeing service which includes an initial advisory service, regular recommendations and advice on optimum routes, and frequent forecasts along the vessel's track, extending to five days ahead. Forecasting and routeing services can be supplied to luxury charter and private yachts of any type. MetWorks services are based on advanced software working with data supplied under contract by the UK Met office.

For further information contact:

Acorn House, Longshot Lane, Bracknell, Berkshire

United Kingdom RG12 1RL

Tel: +44(0)1344 411116 Fax: +44(0)1344 317654

Email: ops@metworksltd.com (FAO Steve Johnson, Director)

TELEPHONE WEATHER SERVICES ABROAD

France

France Telecom and Meteo-France operate an automatic telephone weather service - ALLO METEO FRANCE - with a universal number, the forecast heard depending on the area in which you call as follows:

Service	Telephone number
National forecast	08 92 68 01 01
Regional forecast	08 92 68 00 00
Departmental forecast	08 92 68 02 XX (XX is the local departmental code)
Mountain forecast	08 92 68 04 04
Marine forecast	08 92 68 08 08
Marine local coastal forecast	08 92 68 08 XX (see below)

While the marine forecast service number 08 36 68 08 08 is universally valid and provides a fairly detailed local coastal forecast, a more detailed menu is available.

The following is a selection; prefixed in each case by 08 36 68 08.

Coastal Zone	Telephone number
Pas de Calais	08 92 68 08 59 or 62
Somme estuary	80
Seine estuary	76
Calvados	14
Cotentin	50
Ille et Vilaine	35
Côtes d'Armor	22
Finistère	29
Morbihan	56
Loire-Atlantique	44
Vendée	85
Charente maritime	17

Coastal Zone	Telephone number
Gironde	33
Landes	40
Pyrénées Atlantiques	64
Mediterranean	
Cerbère à Narbonne to The Balearics	08 92 68 08 66 or 11
Narbonne to Port Camargue	30 or 34 or 11
Mouth of the Rhône	30 or 13
Var	83
Alpes Maritimes	06
Corsica, east and west coasts	20

The cost of this service is independent of distance and at a universal rate of 0.34 Euro per minute.

Germany

The Deutscher Wetterdienst provides a telephone weather warnings service which can be accessed from outside the country if required. The number is (49) 40 319 66 28. If there is no warning in operation a wind forecast is given for German Bight and the west and south Baltic.

Ireland

The latest sea area forecast and gale warnings for Irish coasts are available on the Weatherdial Service by calling 1550 123 855.

Spain

The Spanish Instituo Nacional de Meteorologica provides a recorded telephone marine weather information service. Numbers (sea area map on page 63) are:

906 365 371   High seas bulletin for Alboran, Palos, Argelia, Baleares and Golfo de Leon.

                Coastal waters bulletin for Mediterranean coasts.

906 365 372   High seas bulletin for Gran Sol, Vizcaya, Cantabrico, Finisterre.

                Coastal waters bulletin for the coasts of Guipuzcoa, Vizcaya, Cantabria, Asturias, Lugo, Coruña, Pontevedra.

906 365 373   High seas bulletin for Sao Vicente, Azores, Canarias, Sahara, Golfo de Cadiz and Alboran.

                Coastal waters bulletin for coasts of Huelva, Cadiz, Ceuta, Malaga, Melilla, Granada, Almeria and Canary Islands.

906 365 370   Balearic Islands.

NAVTEX (see page 7)

Message category codes:

A – Navigational warning

B – Gale warning

C – Ice report

D – Search and Rescue information

E – Weather forecast

F – Pilot service message

G – Defunct

H – LORAN messages

J – SATNAV messages

K – Other electronic navaid messages

L – Subfacts/Gunfacts Warnings

V – Navigational warnings - additional to A

Z – No messages on hand

Times of NAVTEX weather bulletins

Gale warnings are broadcast by NAVTEX on receipt. Full weather bulletins (A) including 24-hour forecasts for sea areas within 200 to 300 miles of the transmitting station, gale warnings (B) and extended outlooks (C) are broadcast on 518kHz at the following times:

Transmitting station	Times GMT			
Metarea I	(A)		(B) 4-hourly from	(C)
Bodo (B), Norway	0010	1210	0010	
Cullercoats (G), UK	0900	2100	0100	0100
Stockholm (Bjuroklubb)(H), Sweden	0910	2110	0110	
Stockholm (Gislovshammar)(J), Sweden	0930	2130	0130	
Stockholm (Grimeton)(D)	0830	2030	0030	
Rogaland (L), Norway	0150	1350	0150	
Portpatrick (O), UK	0620	1820	0220	0220
Niton (E), UK	0840	2040	0040	2300
Malin Head (Q), Rep of Ireland	1040	2240	0240	
Valentia (W), Rep of Ireland	0740	1940	0340	
Ostend (T), Belgium	0710	1910	0310	
Tallinn (U), Estonia	0720	1920	0320	
Vardo (V), Norway	1130	2330	0330	
Ijmuiden (P)			2030	

Transmitting station	Times GMT			
Metarea II	(A)		(B) 4-hourly from	(C)
Corsen (A), France	0000	1200	0000	
Coruña (D), Spain	0830	2030	0030	
Monsanto (R), Portugal	0905	2105		
Horta (F), Azores			0050	
Tarifa (G) Spain	0900	2100		
Metarea III				
La Garde (W), France (Med coast)	1140	2340	0340	
Valencia (X) Spain	0750	1950		
Split (Q)	4 hourly from 0240		0240	
Cyprus (M)	4-hourly from 0200		0200	
Iraklion (H), Greece	4-hourly from 0110		0110	
Limnos (L), Greece	4-hourly from 0150		0150	
Samsun (E), Turkey	4-hourly from 0030		0030	
Kerkyra (K) Greece	4-hourly from 0140		0140	
Malta (O)	0620	1820		
Rome (R)	0650	1850		
Cagliari (T)	0710	1910		
Augusta (V)	0730	1930		
Trieste (U)	0720	1920		
Haifa (P)	4-hourly from 0020			

NATIONAL NAVTEX SERVICES

A national NAVTEX Service is broadcast on 490kHz twice daily. The schedule at time of going to press was as follows:

		A		B 4-hourly from	Language
Corsen	E	0840	2040		French
Cullercoats	U	0720	1920		English
Horta	J			0930	English
La Garde	S	1100	2300		French
Monsanto	G			0100	
Niton	I	0520	1720		English
Niton	T	0710	1910		French
Portpatrick	C	0820	2020		English

The bulletins include 24-hour **forecasts** and 24-hour **outlooks** (including a forecast of the sea state), and a **three-day national outlook**. Strong-wind warnings are also broadcast on the national NAVTEX transmissions on receipt and then repeated in the routine four-hourly broadcast slots until the warning expires.

FORECASTS FOR EASTERN NORTH ATLANTIC

Radio France Internationale (RFI) provides a full weather bulletin including a 36-hour forecast in French for the Eastern North Atlantic at 1130 GMT daily. The broadcast frequencies are:

6175 kHz for reception in Europe, 11845 for reception in the Mediterranean, 13640, 15300, 21645, 15515, 17570kHz for reception in the Atlantic.

Areas covered are shown on the chart on page 59

SCHEDULE OF WEATHER BULLETINS

Available to yachtsmen in West European waters, transmitted in plain language by radio telephony.

In the following schedule, the references to areas are those indicated on the chart of the appropriate country which follow the schedule. All times are GMT unless indicated otherwise.

The key to the abbreviations used is as follows:

A   Full weather bulletin and forecast.

B   Strong wind warnings, gale warnings and storm warnings.

C   Forecast for coastal waters only.

D   Fog forecast.

1   Time depending on transmitter used.

2   Sea area boundaries as for United Kingdom.

3   Gale warning summaries for appropriate sea areas are broadcast at 0303, 0903, 1503, 2103.

4   See chart for appropriate country.

5   1 April to 30 September only.

6   1 May to 31 August only.

7   Clock times.

8   H + 03, H + 33 until period of validity then change to time given in brackets.

†   Retransmission of earlier broadcast.

*   See also pages 6 to 15.

Notes

In addition to the scheduled services shown, a number of stations broadcast strong wind/gale warnings on receipt and at the end of the next silence period after receipt.

If it is necessary to convert from kHz into metres divide kHz into 300,000 e.g:

200kHz = 300,000 ÷ 200 = 1500 metres

Some broadcasts give wind speed in metres per second. For conversion to knots multiplication by 2 is near enough.

1 knot = 0.515m/sec.   1m/sec = 1.94 knots.

United Kingdom

Forecasts and warnings are broadcast via the HM Coastguard MRCC and MRSC stations. The tables below give the schedule for both the MF and VHF network. All times are clock times.

MF stations (Initial call on 2182 kHz)

Station	Frequency KHz	Shipping Forecast time	Gale, Storm, Strong Wind Warnings†	Sea Areas
Shetland	1770	0905, 2105	0105, 0505, 0905, 1305, 1705, 2105	Faeroes, Fair Isle, Viking
Aberdeen	2226	0720, 1920	0320, 0720, 1120, 1520, 1920, 2320	Fair Isle, Cromarty, Forth, Forties
Humber	2226	0740, 1940	0340, 0740, 1140, 1540, 1940, 2340	Tyne, Dogger, German Bight, Humber
Yarmouth	1869	0840, 2040	0040, 0440, 0840, 1240, 1640, 2040	Humber, Thames
Solent	1641	0840, 2040	0040, 0440, 0840, 1240, 1640, 2040	Portland, Wight
Falmouth	2226	0940, 2140	0140, 0540, 0940, 1340, 1740, 2140	Plymouth, Lundy, Fastnet, Sole
Milford Haven	1767	0735, 1935	0335, 0735, 1135, 1535, 1935, 2335	Lundy, Irish Sea, Fastnet
Holyhead	1880	0635, 1835	0235, 0635, 1035, 1435, 1835, 2235	Irish Sea
Clyde	1883	0820, 2020	0020, 0420, 0820, 1220, 1620, 2020	Bailey, Hebrides, Rockall, Malin
Stornoway	1743	0910, 2110	0110, 0510, 0910, 1310, 1710, 2110	Fair Isle, Faeroes, Hebrides, Bailey, Malin, Rockall

† Note that Gale, Storm or Strong Wind Warnings are broadcast on receipt from the Met Office and repeated in the four-hourly slots indicated while the warning is in force. Strong wind warnings are only broadcast if the warning indicates a significant change to the inshore waters forecast which could result in problems for small craft.

VHF Service (Initial call on Ch16 and then broadcast on Channels 10, 23, 73, 84 or 86)

All UK and International sea areas referred to in the following tables can be found in the relevent charts on pages 55 - 63

Station	Shipping forecast	Local Inshore Waters Forecast Gale, Storm & Strong Wind Warnings†	Sea Areas
Shetland	0905, 2105	0105, 0505, 0905, 1305, 1705, 2105	Faeroes, Fair Isle, Viking
Aberdeen	0720, 1920	0320, 0720, 1120, 1520, 1920, 2320	Fair Isle, Cromarty, Forth, Forties
Forth	1005, 2205	0205, 0605, 1005, 1405, 1805, 2205	Forth, Tyne, Dogger, Forties
Humber	0740, 1940	0340, 0740, 1140, 1540, 1940, 2340	Tyne, Dogger, Humber
Yarmouth	0840, 2040	0040, 0440, 0840, 1240, 1640, 2040	Humber, Thames
Thames	0810, 2010	0010, 0410, 0810, 1210, 1610, 2010	Thames, Dover
Dover	0905, 2105	0105, 0505, 0905, 1305, 1705, 2105	Thames, Dover, Wight
Solent	0840, 2040	0040, 0440, 0840, 1240, 1640, 2040	Portland, Wight
Portland	1020, 2220	0220, 0620, 1020, 1420, 1820, 2220	Plymouth, Portland, Wight
Brixham	0850, 2050	0050, 0450, 0850, 1250, 1650, 2050	Plymouth, Portland
Falmouth	0940, 2140	0140, 0540, 0940, 1340, 1740, 2140	Plymouth, Lundy, Fastnet, Sole
Swansea	0805, 2005	0005, 0405, 0805, 1205, 1605, 2005	Lundy, Irish Sea, Fastnet
Milford Haven	0735, 1935	0335, 0735, 1135, 1535, 1935, 2335	Lundy, Irish Sea, Fastnet
Holyhead	0635, 1835	0235, 0635, 1035, 1435, 1835, 2235	Irish Sea
Liverpool	1010, 2210	0210, 0610, 1010, 1410, 1810, 2210	Irish Sea, Malin
Belfast	0705, 1905	0305, 0705, 1105, 1505, 1905, 2305	Irish Sea, Malin
Clyde	0820, 2020	0020, 0420, 0820, 1220, 1620, 2020	Malin, Hebrides
Oban	0640, 1840	0240, 0640, 1040, 1440, 1840, 2240	Malin, Hebrides
Stornoway	0910, 2110	0110, 0510, 0910, 1310, 1710, 2110	Fair Isle, Malin, Hebrides, Rockall, Bailey, Faeroes

† Note that Gale, Storm or Strong Wind Warnings are broadcast on receipt from the Met Office and repeated in the four-hourly slots indicated while the warning is in force. Strong wind warnings are only broadcast if the warning indicates a significant change to the inshore waters forecast which could result in problems for small craft.

Station	Channel	Frequency kHz	Times of broadcast GMT	Sea areas[4]	Information given	Language
Jersey	25,82	1659	0645[8] 0745[8] 0845[6,8] 1245 1845 2245	Waters around Channel Isles, south of 50°N and east of 3°W	A[7]	English
BBC Radio 4[8]		198, MW and FM	0048 0536 1754	All U.K. sea areas	A	English
		198, and MW	1201	All U.K. sea areas	A	English
		198, MW and FM	0053 0540	U.K. inshore waters	C	English
BBC Radio Guernsey		1116 93.2MHz	0807 1235 (Mon-Fri) 0810 Sat, Sun	Waters around Guernsey	C	English
BBC Radio Jersey		1026 88.8MHz	0635 0810 1835 (Mon-Fri) 0735 Sat, Sun	Waters around Jersey	C	English

Croatia

NAVTEX

Q	Split (Hvar I.)	518 kHz	43°11' N 16°26' E

Weather Bulletins	
Q: 0240 0640 1040 1440 1840 2240	Gale warnings, synopsis and 24 hour forecast in English for Adriatic Sea and Strait of Otranto.

Cyprus

NAVTEX

M	Cyprus	518 kHz	35°02' N 33°17' E

Weather Bulletins	
M: 0200 0600 1000 1400 1800 2200	Gale warnings and forecast in English for Eastern Mediterranean.

Denmark

NAVTEX

D	Tórshavn	518 kHz	62°01' N 6°48' W

Weather Bulletins	
D: 0030 0430 0830 1230 1630 2030	Weather bulletins.

Estonia

NAVTEX

U	Tallinn	518 kHz	59°30' N 24°30' E

Weather Bulletins	
U: 0320 0720 1120 1520 1920 2320	Gale warnings in English for areas B4-B8.
U: 0720 1920	Forecast in English for Baltic Sea.

France (includes Monaco)

NAVTEX (CROSS)

A	Corsen	518 kHz	48°28' N 5°03' W
W	La Garde		43°06' N 5°59' E
E	Corsen	490 kHz	48°28' N 5°03' W
S	La Garde		43°06' N 5°59' E

Weather Bulletins

A: 0000 0400 0800 1200 1600 2000	Storm warnings in English for areas 16-21.
A: 0000 1200	Storm warnings, synopsis and development and 24 hour forecast in English for areas 16-21.
W: 0340 0740 1140 1540 1940 2340	Storm warnings for areas 514 (Eastern part), 515, 516, 521-523, 531-534.
W: 1140 2340	Storm warnings, synopsis and development and 24 hour forecast in English for areas 514 (Eastern part), 515, 516, 521-523, 531-534.
E: 0040 0440 0840 1240 1640 2040	Storm warnings in French for areas 14-26.
E: 0840 2040	Storm warnings, synopsis and development and 24 hour forecast in French for areas 14-29.
S: 0300 0700 1100 1500 1900 2300	Storm warnings in French for areas 514 (Eastern part), 515, 516, 521-523, 531-534.
S: 0700 1900	Storm warnings, synopsis and development and 24 hour forecast in French for areas 514 (Eastern part), 515, 516, 521-523, 531-534.

Germany

NAVTEX

S	Pinneberg	518 kHz	53°43' N 9°55' E
L		490 kHz	

Weather Bulletins

S: 0300 0700 1100 1500 1900 2300	Wind warnings and weather forecast in English for German coastal waters and navigational warning area in the North Sea.
L: 0150 0950 1750	Wind warnings, weather forecast for 12 hours and outlook for a further 12 hours in German for German coastal waters and navigational warning area in the Baltic Sea.
L: 0550 1350 2150	Wind warnings, weather forecast for 12 hours and outlook for a further 12 hours in German for German coastal waters and navigational warning area in the North Sea.

Greece

NAVTEX

H	Irakleio (Iráklion)	518 kHz	35°20' N 25°07' E
K	Kerkyra		39°37' N 19°55' E
L	Limnos		39°52' N 25°04' E

Weather Bulletins

H: 0510 0910 1710 2110	Gale warnings, synopsis, 24 hour forecast and outlook for a further 12 hours in English for areas Saronikós, SE Aegean, SW Aegean, SW Kritiko and SE Kritiko.
K: 0540 0940 1740 2140	Gale warnings, synopsis, 24 hour forecast and outlook for a further 12 hours in English for areas South Adriatic Sea, North / South Ionian Sea, Patraïkós, Korinthiakós and Kithira Sea.
L: 0550 0950 1750 2150	Gale warnings, synopsis, 24 hour forecast and outlook for a further 12 hours in English for areas North / South Aegean, Ikaro, Samos Sea, Saronikós, South Evvoikos, Kafireas Strait, Central / North Aegean, Thrakiko and Thermaïkós.

Ireland

NAVTEX

Q	Malin Head	518 kHz	55°22' N 07°21' W
W	Valentia		51°56' N 10°21' W

Weather Bulletins

Q: 0240 0640 1040 1440 1840 2240	Gale warnings for Irish coastal waters up to 30 n miles offshore and areas Irish Sea, Shannon, Rockall, Malin and Bailey.
Q: 0640 1840	Gale warnings and forecast for areas Shannon, Rockall, Malin and Bailey.
Q: 1040 2240	Gale warnings, synopsis and 24 hour forecast for Irish coastal waters up to 30 n miles offshore and areas Irish Sea, Atlantic-East Northern and East Central Sections.
W: 0340 0740 1140 1540 1940 2340	Gale warnings for Irish coastal waters up to 30 n miles offshore and area Irish Sea.
W: 0740 1940	Gale warnings, synopsis and 24 hour forecast for Irish coastal waters up to 30 n miles offshore and areas Irish Sea, Sole, Fastnet and Shannon.
W: 1140 2340	Forecast for area Atlantic-East Central Section.

Italy

NAVTEX

V	Augusta	518 kHz	37°14' N 15°14' E
T	Cagliari		39°14' N 09°14' E
R	Roma		41°48' N 12°31' E
U	Trieste		45°41' N 13°46' E

Weather Bulletins

V: 0330 0730 1130 1530 1930 2330	Storm warnings.
V: 0730 1930	Weather forecasts.
T: 0310 0710 1110 1510 1910 2310	Storm warnings.
T: 0710 1910	Weather forecasts.
R: 0250 0650 1050 1450 1850 2250	Storm warnings
R: 0650 1850	Weather forecasts.
U: 0320 0720 1120 1520 1920 2320	Storm warnings.
U: 0720 1920	Weather forecasts.

Malta

NAVTEX

O	Malta	518 kHz	35°49' N 14°32' E

Weather Bulletins

O: 0620 1820	Gale warnings, synopsis, 12 hour forecast for coastal waters of Malta up to 50 n miles offshore.

Netherlands

NAVTEX Netherlands Coastguard

P	Den Helder	518 kHz	52°57' N 4°48' E

Weather Bulletins

P: 0230 0630 1030 1430 1830 2230	Gale warnings in English for areas, Thames, Humber, German Bight and Dogger.
P: 0230 1430	Forecasts in English for areas Viking, Forties, Dogger, Fisher, German Bight, Humber, Thames and Dover.

Norway

B	Bodø	518 kHz	67°16' N 14°23' E
N	Ørlandet		63°41' N 09°35' E
L	Rogaland		58 °39' N 05°36' E
V	Vardø		70°22' N 31°06' E
A	Svalbard		78°04' N 13°37' E

Weather Bulletins

B: 0010 1210	Weather forecast in English for coastal waters off northern Norway (65°N-70°N, west to 5°W) Norwegian and Arctic Sea (north of 70°N from 20°E to 5°W).
N: 0210 1410	Weather forecast in English for coastal waters off Norway (62°N-65°N, west to 5°W).
L: 0150 1350	Weather forecast in English for North Sea, Skagerrak and coastal waters off southern and western Norway (north to 62°N, west to 0°W).
V: 1130 2330	Weather forecast in English for coastal waters off northern Norway (east of 20°E) Arctic and Barents Sea (east of 20°E).
A: 0000 1200	Weather forecast in English for coastal waters off northern Norway (65°N to 70°N, west to 5°W), Norwegian and Arctic Sea (north of 70°N) and Barents Sea.
A: On request	Weather warnings.

Portugal

R	Monsanto (Alges)	518 kHz	38°44' N 9°11' W
G		490 kHz	

Weather Bulletins

R: 0250 0650 1050 1450 1850 2250	Storm warnings and Gale warnings, synopsis and 24 hour forecast in English for areas Charcot, Josephine, Finisterre, Porto, S. Vicente and Cádiz.
G: 0100 0500 0900 1300 1700 2100	Storm and gale warnings, synopsis and 24 hour forecast in Portuguese for areas 4, 6 and 16-19.

Spain (includes Gibraltar)

NAVTEX

D	Coruña	518 kHz	43°22' N 8°27' W
G	Tarifa		36°01' N 5°35' W
X	Valencia (Cabo de la Nao)		38°43' N 0°09' E
W	Coruña	490 kHz	43°22' N 8°27' W
T	Tarifa		36°01' N 5°35' W
M	Valencia (Cabo de la Nao)		38°43' N 0°09' E

Weather Bulletins

D: 0830 2030	Storm warnings, synoptic situation and development and forecast in English valid for 24 hours for N Atlantic within 450 n miles of coast.
G: 0900 2100	Storm warnings, synoptic situation and development and forecast in English valid for 24 hours for N Atlantic areas San Vicente, Cadiz and Casablanca and West Mediterranean Sea within 450 n miles of coast.
X: 0750 1950	Storm warnings, synoptic situation and development and forecast in English valid for 24 hours for West Mediterranean Sea within 450 n miles of coast.
W: 1140 1940	Storm warnings, synoptic situation and development and forecast in Spanish valid for 24 hours for N Atlantic within 450 n miles of coast.
T: 0710 1910	Storm warnings, synoptic situation and development and forecast in Spanish valid for 24 hours for N Atlantic areas San Vicente, Cadiz and Casablanca and West Mediterranean Sea within 450 n miles of coast.
M: 1000 1800	Storm warnings, synoptic situation and development and forecast in Spanish valid for 24 hours for West Mediterranean Sea within 450 n miles of coast.

Sweden
(all Navtex broadcasts in the Baltic are co-ordinated by Sweden)

NAVTEX

H	Bjuröklubb	518 kHz	64°28' N 21°36' E
J	Gislövshammar		55°29' N 14°19' E
I	Grimeton		57°06' N 12°23' E

Weather Bulletins

H: 0110 0510 0910 1310 1710 2110	Gale warnings in English for areas B1-B3.
H: 0510 1710	Forecast in English for Baltic Sea.
I: 0120 0520 0920 1320 1720 2120	Gale warnings in English for areas B12-B14.
I: 0520 1720	Forecast in English for Baltic Sea.
J: 0130 0530 0930 1330 1730 2130	Gale warnings in English for areas B9-B11.
J: 0530 1730	Forecast in English for Baltic Sea.

Turkey

NAVTEX

F	Antalya (Mediterranean Coast)	518 kHz	36°53' N 30°42' E
D	Istanbul (Marmara Denizi)		41°04' N 28°57' E
I	IzmÝr (Aegean Coast)		38°17' N 26°16' E
E	Samsun (Black Sea Coast)		41°17' N 36°20' E
D	Antalya (Mediterranean Coast)	490 kHz	36°53' N 30°42' E
B	Istanbul (Marmara Denizi)		41°04' N 28°57' E
C	IzmÝr (Aegean Coast)		38°17' N 26°16' E
A	Samsun (Black Sea Coast)		41°17' N 36°20' E
M	Istanbul (Marmara Denizi, Aegean, Mediterranean and Black Sea)	4209.5 kHz	41°04' N 28°57' E

Weather Bulletins

D: 0030 0430 0830 1230 1630 2030	Weather forecast in English for areas Marmara and Danube.
E: 0040 0440 0840 1240 1640 2040	Weather forecast in English for areas Georgia and Danube.
F: 0050 0450 0850 1250 1650 2050	Weather forecast in English for area Taurus.
I: 0120 0520 0920 1320 1720 2120	Weather forecast in English for areas Aegean and Jason.
A: 0000 0400 0800 1200 1600 2000	Weather forecast in Turkish for areas Georgia and Danube.
B: 0010 0410 0810 1210 1610 2010	Weather forecast in Turkish for areas Marmara and Danube.
C: 0020 0420 0820 1220 1620 2020	Weather forecast in Turkish for areas Aegean and Jason.
D: 0030 0430 0830 1230 1630 2030	Weather forecast in Turkish for area Taurus.
M: 0200 0600 1000 1400 1800 2200	Weather forecast in Turkish for areas 12-27.

UNITED KINGDOM

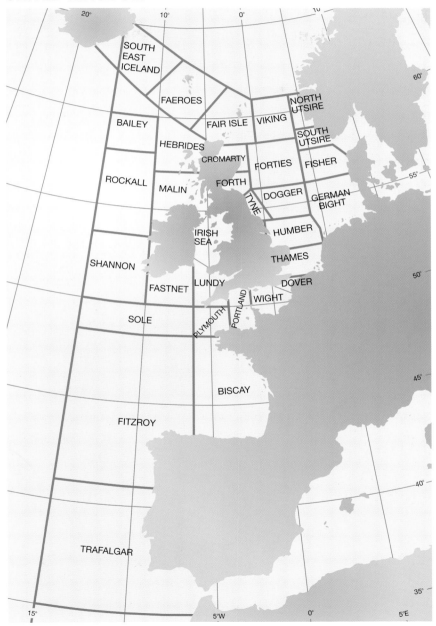

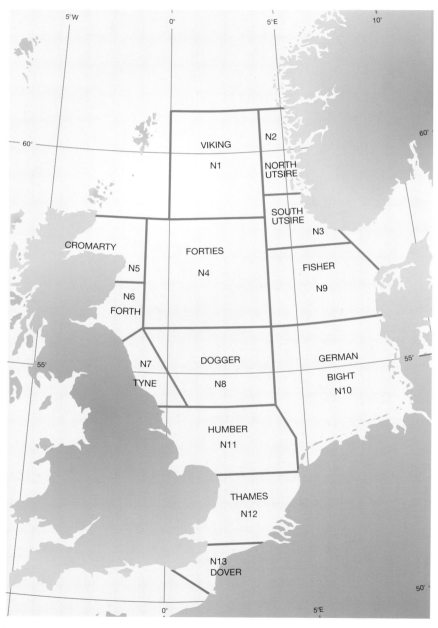

BALTIC COMMON AREAS

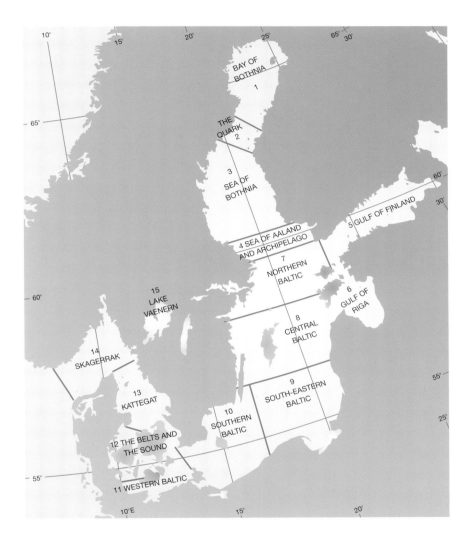

The Mediterranean areas above are also used on Inmarsat - C, and French and Spanish Navtex.

FRANCE

Sea areas east of 35°W used by Radio France International (see page 43)

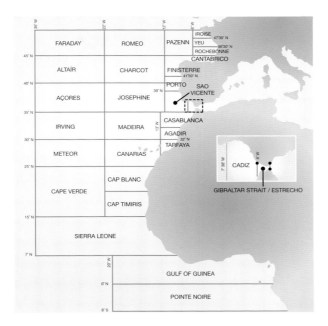

FRANCE

CROSS areas of responsibility (see page 48)

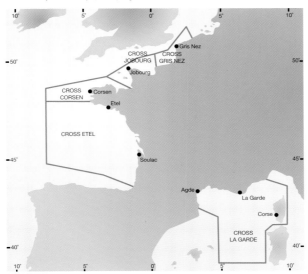

FRANCE / SPAIN / PORTUGAL

Metarea II sea areas used in GMDSS and Navtex (See pages 6 & 42)

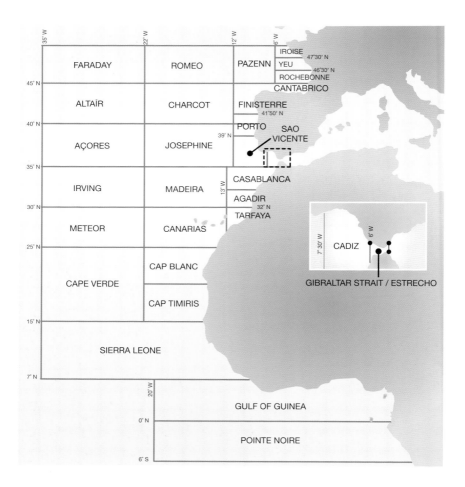

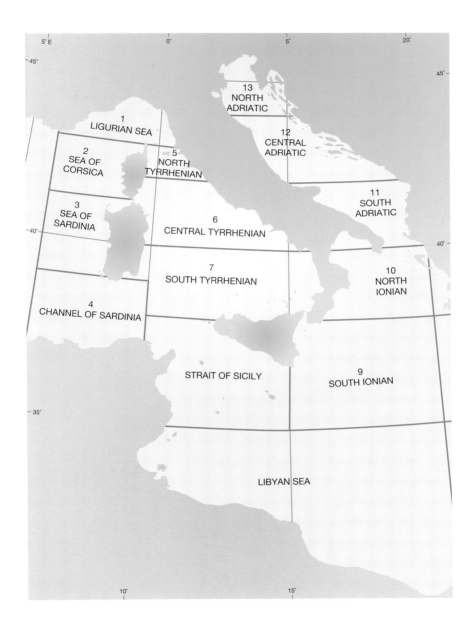

5° E   0°   5°   20°

45°   45°

13 NORTH ADRIATIC

1 LIGURIAN SEA

12 CENTRAL ADRIATIC

2 SEA OF CORSICA

5 NORTH TYRRHENIAN

11 SOUTH ADRIATIC

3 SEA OF SARDINIA

40°   **6 CENTRAL TYRRHENIAN**   40°

7 SOUTH TYRRHENIAN

10 NORTH IONIAN

4 CHANNEL OF SARDINIA

STRAIT OF SICILY

9 SOUTH IONIAN

35°

LIBYAN SEA

10°   15°

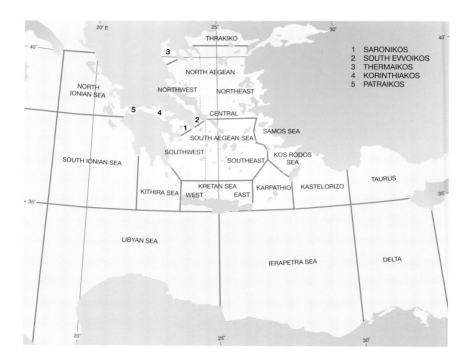

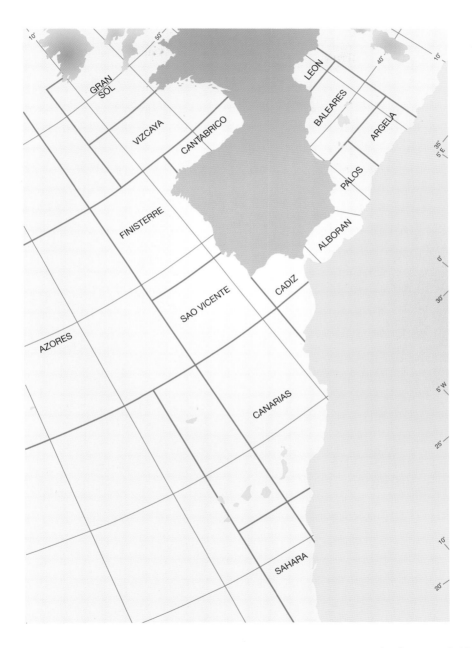

International weather

ENGLISH	DANISH	DUTCH	FRENCH	GERMAN	ITALIAN	SPANISH
A Amendment	Aendring	Verandering	Changement Amendement	Änderung	Correzione	Enmienda Rectification
Area	Farvand	Gebied	Zone	Gebiet	Area	Zona
B Backing	Venstredrejende	Krimpend	Retournant	Rückdrehend	Rotazione a Sinistra Rotazione Antioraria	Rolada a la Izquierda
Beaufort Wind Scale	Beaufort's Vindskala	Beaufortschaal voor Windkracht	Echelle de Beaufort	Beaufortskala	Scala di Beaufort	Escala Beaufort
C Calm	Vindstille	Windstilte	Calme	Windstille	Calmo	Calma
Centre	Centrum Center	Centrum	Centre	Zentrum	Centro	Centro
Choppy	Skiftende	Woelig	Hachée	Kabbelige	Mosso	Agitado
Clouds	Skyer	Wolken	Nuages	Wolken	Nubi	Nubes
Clouds (broken)	Skyet	Gebroken	Nuages Fragmentés Troué	Bewölkung	Nubirotte	Quebrado, Nubes Fragmentadas
Cloudy	Overskyet	Bewolkt	Nuageux	Bewölkt	Nuvoloso	Nublado, Nuboso
Coast	Kyst	Kust	Côte	Küste	Costa	Costa
Coastal	Kyst-	Kust-	Littoral	An der Küste	Costiero	Costero
Cold	Kold	Koud	Froid	Kalt	Freddo	Frio
Cyclonic	Cyklonisk	Cycloonachtig, Cyclonisch	Cyclonique	Zyklonisch	Ciclonico	Ciclonica
D Dawn	Daggry	Dageraad Morgenschem-ering	Aube, au point du jour	Morgen-dämmerung	Alba	Alba
Decrease (Wind)	Aftagen	Afnemen	Affaiblissement, Diminution	Abnahme	Caduto, Diminuzione	Disminución
Deep	Dyb	Diep	Profond	Tief	Profondo	Profundo
Deepening	Uddybende	Verdiepend	Creusement	Vertiefung	Approfondi-mento	Ahondamiento
Dense	Taet, Tyk	Dicht	Dense	Dicht	Denso	Denso
Depression (Low)	Lavtryk	Depressie	Dépression	Tief	Depressione	Depresión
Direction	Retning	Richting	Direction	Richtung	Direzione	Dirreción
Dispersing	Som spreder sig	Verstrooiend	Se dispersant, se dissipant	Zerstreuung	Dispersione	Disipación
Drizzle	Finregn	Motregen	Crachins	Sprühregen	Spruzzatore, Pioviggne	Lloviznaa
Dusk	Tusmørke	Avondschem-ering	Brune, crépuscule du soir	Abend-dämmerung	Crepusculo Tramontana	Crepúsculo

ENGLISH	DANISH	DUTCH	FRENCH	GERMAN	ITALIAN	SPANISH
E East	Øst	Oosten	Est	Ost	Est Levante	Este
Extending	Udstraekkende	Uitstrekkend	Extension	Ausbreitend	Estendo	Extension
Extensive (or Widespread)	Udstrackt	Uitgestrekt	Extendue	Verbreitet	Esteso	General
F Falling	Faldende	Dalend, Vallend	En Baisse	Fallend	In diminuzione	En disminucion
Filling	Udfyldende	(Op) Vullend	Comblement	Auffüllend	Riempimento Colmamento	Relleno
Fine (or Fair)	Smukt, (Klart)	Mooi	Clair, Beau	Schönwetter	Sereno, bello	Sereno
Fog	Taage	Mist	Brouillard	Nebel	Nebbia	Niebla
Fog bank	Taage Banke	Mist Bank	Banc de Brouillard	Nebelbank	Banco di Nebbia	Banco de Niebla
Forecast	Vejrforudsigelse	Verwachting	Prévision	Vorhersage	Previsione	Previsión
Formation	Formation	Formatie	Formation	Bildung	Formazione	Formación
Forming	Danne (dannende)	Vorming	Developpent,	Formend,	Formando se formant	Formante bildend
Frequent	Hyppig	Veelvuldig	Fréquent	Häufig	Frequente	Frecuenta
Fresh	Frisk	Fris	Fraiche, frais	Frisch	Fresco	Fresco
Front	Front	Front	Front	Front	Fronte	Frente
Front (passage of)	Front passage	Front Passage	Passage d'un Front	Frontdurch- gang	Passaggio di un Fronte	Paso de un Frente
Frost	Frost	Vorst	Gelée	Frost	Brina	Escarcha
G Gale	Stormende Kuling, Hard Kuling	Stormachtig	Coup de Vent	Stürmischer wind	Burrasca	Viento Duro
Cones (Gale)	Oje	Kegel	Cône	Sturmekegel	Sintoma	Cona
Gale warning	Stormvarsel	Stormwaarsch- uwing	Avis de coup de Vent	Sturmwarnung	Avviso di Burrasca	Aviso de Temporal
Good	God	Goed	Bon	Gut	Buono	Bueono
Gust	Vinstød, Vindkast	Windstoot	Rafale	Windstoss	Colpo di Vento, Raffica	Ráfaga, Racha
Gusty	Stormfuld, Byget	Buiig	(Vent) à Rafales	Böig	Con Raffiche	en Räfagas en Rachas
H Hail	Hagl	Hagel	Grêle	Hagel	Grandine	Granizo
Haze	Dls	Nevel	Brume Sèche	Dunst (Trockener)	Caligine	Calina
Hazy	Diset	Nevelig	Brumeux	Diesig	Caliginoso	Calinoso
Heavy	Svaer, Kraftig	Zwaar	Abondant, Violent	Ergiebig, Schwer	Pesante, Violento	Abunante, Violento
High (Anticyclone)	Anticyklon, Højtryk	Hogedruk- gebeid	Anticyclone	Antizyklone Hockdruck- gebiet	Anticiclone	Anticiclón

ENGLISH	DANISH	DUTCH	FRENCH	GERMAN	ITALIAN	SPANISH
Hurricane	Orkan	Orkaan	Ouragan	Orkan (auuserhalb der Tropen), Hurrikan	Uragano	Huracán
I Increasing	Tiltagende, øgende	Toenemend	Augmentant	Zunehmend	In Auhmento	Aumentar
Intermittent	Intermitternde, Tiltider, Tidvis	Afwisselend	Intermittent	Zeitweilig	Intermittente	Intermitente
Isobar	Isobar	Isobar	Isobare	Isobare	Isobara	Isobara
Isolated	Isolere, Enkelte	Verspreid	Isolé	Einzelne	Isolato	Aislado
L Latitude	Bredde	Breedte	Latitude	Breite	Latitudine	Latitud
Light, slight	Tynd, let	Licht, Gering, Zwak	Faible	Schwach	Leggero, Debole	Ligero, Dehil
Lightning	Lyn	Bliksem	Eclair Foudre	Blitz	Lampo	Recámpago
Line squall	Bygelinie	Buienlijn	Ligne de Grain	Böenfront	Linea di Groppo	Linea de Turbonada
Local	Lokal	Plaatselijk	Locale	Örtlich, Lokal	Locale	Local
Longtitude	Laengde	Lengte	Longitude	Länge	Longitudine	Longitud
M Meridian	Meridian	Meridiaan	Méridien	Meridian Langenkreis	Meridiano	Meridiano
Mist	Let Tåge, Tågedis	Nevel	Brume Légère	Dunst (Feuchter)	Foschia, Brumo	Neblina
Misty	Taget, Diset	Nevelig	Brumeux	Dunstig, Diesig	Brumoso,	Fosco Brumoso
Moderate	Middlemadig, Moderat	Matig, Gematigd	Modéré	Mässig	Moderato	Moderado
Moderating	Beherske	Matigend, Afnemend	Se Modérant Se Calmant	Abschwäch-end Abnehmend	Medianente, Calmante	Medianente
Morning (in the)	Om Formiddagen Om Morgenen	Morgen, Voormiddag	Le Matin	Morgens	Al Mattino, Par il Mattino	Por la Manana
Moving	Bevaegende	Bewegend	Se Déplaçant	Ziehend	In Movimento, Si Muove	Movimiento
N North	Nord	Noorden	Nord	Nord	Settentionale, Nord	Septentrional Boreal
O Occasional	Af og til, Ti tider	At en toe	Eparses	Teilweise	Occasionale	Occasional
Occlusion	Okklusion	Okklusie	Occlusion	Okklusion	Occlusion	Oclusion
Off-shore wind	Fra land vind	Aflandige wind	Vent de Terre	Ablandiger Wind, Landwind	Vento (Brezza) di Terra	Viento Terral
On-shore wind	I land vind Palånsvind	Wind van zee	Vent de mer Brise de mer	Auflandiger Wind, Seewind	Vento (Brezza) di mare	Viento demar
Overcast	Overtrukket Overskyet	Geheel bewolkt	Couvert	Bedeckt	Coperto	Cubierto

ENGLISH	DANISH	DUTCH	FRENCH	GERMAN	ITALIAN	SPANISH
P Period	Periode	Tijdvak, Periode	Période	Periode	Periodo	Periodo
Period of validity	Glydigheds-periode	Geldigheids-duur	Période de Validité	Gültigkeits-dauer	Periodo di Validità	Periodo de Validez
Poor	Ringe, Sigt	Gering Slecht	Mauvais	Schlecht	Mao, Scarso	Mal
Precipitation	Nedbør	Neerslag	Précipitation	Niederschlag	Precipitazione	Precipitación
Pressure	Tryk	Druk	Pression	Druck	Pressione	Presión
Q Quickly	Kvick, Hurtigt	Zeer Snel	Rapidement	Schnell	Pronto	Pronto
R Rain	Regn	Regen	Pluie	Regen	Pioggia	Pluvial, Lluvia
Continuous (rain)	Uafbrudt redvarende	Onafgebroken	Continue	Anhaltend	Continua	Continuo
Slight (rain)	Let	Licht, Gering	Faible	Leicht	Pioggia debole	Débil Legero
Ridge	Ryg	Rug	Dorsale	Rücken	Promontorio	Dorsal
Rising	Stigning	Rijzend, Stijgend	En Hausse	Steigend	Ascendente	Ascendente
Rough	Oprørt	Guur	Agitée	Stürmisch	Agitato, Grosso	Bravo o alborotado
S Scattered	Spredt, Stro	Verspreide	Sporadiques	Zerstreut	Diffuso	Difuso
Sea	Sø, Hav	Zee	Mer	Meer	Mare	Mar
Sea breeze	Søbrise, Havbris	Zeewind	Brise de Mer	Seebrise	Brezze di mare	Virazón
Shower	Byge	Regenbui	Averse	Regenschauer	Rovescio	Aguacero, Chubasco
Sleet	Slud, sne og regne	Natte sneeuw	Grésil	Schneeregen	Nevischio	Aguanieve
Slowly	Langsomt	Zie Langzaam	Lentement	Langsam	Lentamente	Lentamente
Smooth	Glatte	Vlak	Belle	Glatt	Tranquillo, Calmo	Tranquilo, Calmo
Snow	Sne	Sneeuw	Neige	Schnee	Neve	Neive
South	Syd	Zuiden	Sud	Süd	Meridionale	Sur
Squall	Byge	Windvlaag	Grain	Böe	Tempestra	Turbonada
State of sea	Sø Stat, Søeus Tilstand	Toestand van de zee	État de la mer	Zustand der See	Stato del mare	Estado del mare
Stationary	Stationaer	Stationair	Stationnaire	Stationär	Stazionario	Estacionario
Steadily	Regelmaessig	Geregeld, Regelmatig	Regulièrement	Regelmässig	Constante-mente	Constante-mente
Storm	Uneir	Storm	Tempête	Sturm	Tempesta, Tembrale	Temporal
Strong	Staerk, Kraftig	Sterk, Krachtig	Fort	Stark	Forte	Fuerte
Swell	Dønning	Deining	Houle	Dünung	Onda lunga, Mare lungo	Martendida, Mar de Leva
T Thunder	Torden	Donder	Tonnerre	Donner	Tuono	Tormenta
Thunderstorm	Tordenvejr	Onweer	Orage	Gewitter	Temporale	Trueno
Time	Tid	Tijd	Temps	Zeit	Tempo	Hora
Trough	Udlober, Trug	Trog	Creux	Trog	Saccatura	Vaguada

ENGLISH	DANISH	DUTCH	FRENCH	GERMAN	ITALIAN	SPANISH
V Variable	Foranderlig Variabel	Veranderliik	Variable	Veränderlich	Variabile	Variable
Veering	Drejer til Højre	Ruimend	Virement ou Virage	Rechtsdrehend, Ausschiessen	Rotazione ovaria, Rotazione a destra	Dextrogiro
Visibility	Sigt, Sigtbarhed	Zicht	Visibilité	Sicht	Visibilità	Visibilidad
W Warm	Varm	Warm	Chaud	Warm	Cáldo	Cálido
Waterspout	Skypumpe	Waterhoos	Trombe Marine	Wasserhose	Tromba Marina	Tromba Marina
Wave formation	Bølgeformation	Golfformatie	Formation des Vagues	Wellenbildung	Formazione di onde	Formación de ondas
Weather	Vejr	Weer	Temps	Wetter	Tempo	Tiempo
Weather conditions	Vejr-Betingelse	Weer-somstandig-heden	Conditions du Temps	Wetter-Verhältnisse	Condizioni Tempo	Condiciones del Tiempo
Weather report	Vejrmelding	Weerbericht	Rapport, Météorol-ogique	Wettermel-dung Wetterbericht	Rapporto, (Bollentino) Meteoroligico	Informe, aviso Boletin Meteorologico
West	Vest	Westen	Ouest	West	Ovest, Ponente	Oeste
Whirlwind	Hvirvelvind	Windhoos	Tourbillon de Vent	Wirbelwind	Turbine	Torbellino
Wind	Vind	Wind	Vent	Wind	Vento	Viento
Wind Force	Vindstyrke	Windkracht	Force du Vent	Windstärke	Forza (Intensita) del Vento	Intensidad Fuerrza del Viento

NUMBERS

One	Een	Een	Un(e)	Eins	Uno	Uno
Two	To	Twee	Deux	Zwei	Due	Dos
Three	Tre	Drie	Trois	Drei	Tre	Tres
Four	Fire	Vier	Quatre	Vier	Quattro	Cuatro
Five	Fem	Vijf	Cinq	Fünf	Cinque	Cinco
Six	Seks	Zes	Six	Sechs	Sei	Seis
Seven	Syv	Zeven	Sept	Sieben	Sette	Siete
Eight	Otte	Acht	Huit	Acht	Otto	Ocho
Nine	Ni	Negen	Neuf	Neun	Nove	Nueve
Ten	Ti	Tien	Dix	Zehn	Dieci	Diez

Index

Index

Index

RYA Weather Handbook

To feel safe and confident at sea, we all want to know what is likely to happen weather-wise over the next few hours and days ahead.

The RYA Weather Handbook has been written to complement the RYA Dayskipper and Coastal Skipper/Yachtmaster syllabus and is recommended reading for anyone taking these courses.

But whether you are following a RYA course or just want to understand the weather a little better, this book is full of practical and useful advice on how to understand weather maps and improve your forecasting skills.

Written by a sailor for sailors, the RYA Weather Handbook explains the complexities of the weather using full colour diagrams and illustrations - and without lots of jargon.

Your time afloat should be fun and understanding the weather can only enhance your enjoyment.

Ref: G1

RYA *Membership*

Promoting and Protecting Boating
www.rya.org.uk

1 **Important** To help us comply with Data Protection legislation, please tick *either* Box A or Box B (you must tick Box A to ensure you receive the full benefits of RYA membership). The RYA will not pass your data to third parties.

☐ A. I wish to join the RYA and receive future information on member services, benefits (as listed in RYA Magazine and website) and offers.

☐ B. I wish to join the RYA but do not wish to receive future information on member services, benefits (as listed in RYA Magazine and website) and offers.

When completed, please send this form to: RYA, RYA House, Ensign Way, Hamble, Southampton, SO31 4YA

2

	Title	Forename	Surname	Date of Birth	Male	Female
1.				DD / MM / YY	☐	☐
2.				DD / MM / YY	☐	☐
3.				DD / MM / YY	☐	☐
4.				DD / MM / YY	☐	☐

Address

Town **County** **Post Code**

Evening Telephone **Daytime Telephone**

email

Signature: **Date:**

3 **Type of membership required:** *(Tick Box)*

☐ *Personal* Annual rate £39 or £36 by Direct Debit

☐ *Under 21* Annual rate £13 (no reduction for Direct Debit)

☐ *Family** Annual rate £58 or £55 by Direct Debit

* Family Membership: 2 adults plus any under 21s all living at the same address

4 Please tick ONE box to show your main boating interest.

☐ Yacht Racing
☐ Dinghy Racing
☐ Personal Watercraft
☐ Powerboat Racing
☐ Motor Boating
☐ Yacht Cruising
☐ Dinghy Cruising
☐ Inland Waterways
☐ Windsurfing
☐ Sportsboats and RIBS

Please see Direct Debit form overleaf

Instructions to your Bank or Building Society to pay by Direct Debit

Please complete this form and return it to:
Royal Yachting Association, RYA House, Ensign Way, Hamble, Southampton, Hampshire SO31 4YA

Originators Identification Number

9	5	5	2	1	3

To The Manager: _____ Bank/Building Society

Address: _____

_____

Post Code: _____

2. Name(s) of account holder(s)

_____

_____

3. Branch Sort Code

4. Bank or Building Society account number

5. RYA Membership Number (For office use only)

_____

6. Instruction to pay your Bank or Building Society

Please pay Royal Yachting Association Direct Debits from the account detailed in this instruction subject to the safeguards assured by The Direct Debit Guarantee.

I understand that this instruction may remain with the Royal Yachting Association and, if so, details will be passed electronically to my Bank/Building Society.

Signature(s) _____

Date _____

Banks and Building Societies may not accept Direct Debit instructions for some types of account

Cash, Cheque, Postal Order enclosed £ _____

Made payable to the Royal Yachting Association

Office use only: Membership Number Allocated _____

077

Office use / Centre Stamp

Notes

Notes

Notes

Notes